COLLEGE
MATHEMATICS

Kaj L. Nielsen

BARNES & NOBLE, Inc. · New York

Publishers · Booksellers · Since 1873

Dedicated to Janice Ann

This book is an original work (No. 105) in the original College Outline Series. It was written by a distinguished educator, carefully
edited, and manufactured in the U. S. A. in accordance with the
highest standards of publishing.

PREFACE

This outline presents the fundamentals of a first-year course in college mathematics. Such a general course usually is intended for students who do not plan to continue the study of mathematics, but who, nevertheless, would like to obtain a knowledge of the basic concepts of mathematics on a college level. This book will furnish a rapid review for the reader who has completed the course, and will furnish supplementary reading for one who is enrolled in this course of study.

A course in general college mathematics may encompass many things. Each instructor and each university have points of particular emphasis. This book has concentrated on the fundamental concepts of algebra, trigonometry, and analytic geometry, with an introduction to the calculus. The book is complete in itself and, although it is not intended to be used as a textbook, it may be read with profit by those who are beginning the study of college mathematics. It should provide any reader with a thorough knowledge of the usual topics studied during the first year in this field.

Each concept has been illustrated with figures, solved problems, and examples. Neatness and systematic manner of solving the problems have been stressed, for these should be products of the study of mathematics. Logical thinking and elimination of errors can be achieved by the adaptation of systematic procedures. Abbreviated tables have been placed in the back of the book for computational purposes. If the reader finds a need for greater accuracy, he can obtain more extensive tables in this College Outline Series and will not experience any difficulty in using them.

A thorough knowledge of any branch of mathematics cannot be obtained without solving problems. Consequently, a few typical exercises have been placed at the end of each chapter. The student is urged to work all the exercises. Answers are furnished to those that require answers, so that the reader may check his work. A set of examinations, with answers, forms a part of this book, and it is hoped that they may aid the student in his preparation for any examination.

Although this book contains the fundamentals of a first-year college mathematics course, it must be realized that it is in the nature of a survey and that there are many things in each subject matter which could receive greater attention. It is hoped that the readers of this book will not consider this as the end of their study of mathematics, but rather as source material which will lead into other mathematical pursuits.

The author gratefully acknowledges his indebtedness to his many students who, during his teaching career, contributed so much to the manner of presentation of college mathematics; to the scholars who developed the subject matter; to his friends who aided in the preparation of the manuscript; to Dr. Gladys Walterhouse for her careful reading and checking of the manuscript; and to the staff of Barnes & Noble, Inc., who contributed to the finished product.

KAJ L. NIELSEN

TABLE OF CONTENTS

TABULATED BIBLIOGRAPHY
OF STANDARD TEXTBOOKS

This *College Outline* is keyed to standard textbooks in two ways.

1. If you are studying one of the following textbooks, consult the cross references here listed to find which pages of the *Outline* summarize the appropriate chapter of your text. (Roman numerals refer to the textbook chapters, Arabic figures to the corresponding *Outline* pages.)

2. If you are using the *Outline* as your basis for study and need a fuller treatment of a topic, consult the pages of any of the standard textbooks as indicated in the Quick Reference Table on pages xiv-xix.

Allendoerfer, Oakley, *Fundamentals of Freshman Mathematics*, 2nd ed., 1965, McGraw-Hill.

I (232–235); II (1–5, 21–28); III (5–13); IV (13–19); V (7–10); VI (47–50, 100–103, 232–237); VII (79–97); VIII (71–74); IX (21–26); X (155–166); XI (167–169); XII (28–45, 180–188); XIII (28–62); XIV (24–28, 79–144); XV (225–230); XVI (212–224); XVII (none).

Allendoerfer, Oakley, *Principles of Mathematics*, 1963, McGraw-Hill.

I (232–235); II (1–20); III (69–71, 11–13); IV (none); V (23–26, 49–50, 71–73, 79–144, 155–166); VI (21–27, 235–237); VII (167–179); VIII (28–46, 51–63, 74–78); IX (79–89, 115–144); X (145–150, 212–214); XI (212–231); XII (150–154, 66–71); XIII (none).

Apostle, *A Survey of Basic Mathematics*, 1960, Little, Brown.

I (1–2); II (2–3); III (18); IV (4–12); V (12–13); VI (13–16); VII (47–50, 89–98); VIII (7–11); IX (100–114); X–XV (none); XVI (150–154); XVII (21–26); XVIII (none); XIX (167–173); XX (30–38, 180–196); XXI (196–197); XXII (145–150); XXIII (21–27, 79–89, 109–136, 201–211); XXIV (212–231).

Ayres, Fry, Jonah, *General College Mathematics*, 2nd ed., 1960, McGraw-Hill.

I (232–248); II (64–66); III (21–28, 232–237); IV (79–98); V (47–51); VI (100–113, 115–120, 128–130); VII (none); VIII (30–32); IX (32–38, 180–185); X (185–196); XI (28–29); XII (40–45); XIII (none); XIV (167–179); XV (none); XVI (150–154); XVII–XIX (none); XX–XXIV (232–248).

Banks, *Elements of Mathematics*, 1961, Allyn and Bacon.

I (1–3, 232–235); II (none); III (1–6, 237–240); IV (1–6); V (1–19); VI (21–33); VII (1–19); VIII (28–30, 189); IX (167–177); X (66, 146–148); XI (79–136); XII (21–62).

Brink, *A First Year of College Mathematics*, 2nd ed., 1954, Appleton.

I (1–19); II (21–28); III (21–27, 235); IV (79–82); V (82–96); VI (64–66); VII (100–104); VIII (71–74); IX (21–27, 221–225); X (104–109); XI (145–150); XII (66–71); XIII (69–71); XIV (167–179); XV (28–32); XVI (32–37, 180–185); XVII (35–45); XVIII (47–54); XIX

(54–62) ; XX (185–199) ; XXI (40–45) ; XXII (137–144) ; XXIII (74–77) ; XXIV (155–166) ; XXV (82–94) ; XXVI (24–26) ; XXVII (130–134) ; XXVIII (109–113) ; XXIX (221–225) ; XXX (115–135) ; XXXI (150–153) ; XXXII (153–154) ; XXXIII (none) ; XXXIV (241–243) ; XXXV (148–150) ; XXXVI–XXXVII (none) ; XXXVII (201–211) ; Appendix (1–19, 49–51).

Brixey, Andree, *Fundamentals of College Mathematics*, 1961, Holt, Rinehart, Winston.
I (232–237, 1–18, 71–73) ; II (21–27, 71–73) ; III (21–27) ; IV (155–162) ; V (79–90, 22–23) ; VI (212–225) ; VII (221–224) ; VIII (7–10, 167–179) ; IX (177–179) ; X (145–150) ; XI (151–154) ; XII (none) ; XIII (none) ; XIV (28–37, 185–187) ; XV (38–45) ; XVI (45–61, 217) ; XVII (137–144) ; XVIII (75–77) ; XIX (212–231) ; XX (79–99) ; XXI (109–136) ; XXII (none) ; XXIII (none) ; XXIV (201–211) ; XXV (232–248).

Denbow, Goedicke, *Foundations of Mathematics*, 1959, Harper.
I (1–5) ; II (5–19) ; III (1–19) ; IV (21–46) ; V (none) ; VI (89–99) ; VII (18, 74–78, 100–114, 145–152) ; VIII (232–237) ; IX (152–154) ; X (none) ; XI (47–51) ; XII (28–46, 51–63) ; XIII (23–27, 79–91, 109–144) ; XIV (212–231) ; XV–XVI (none) ; XVII (232–248).

Elliott, Miles, *College Mathematics*, 2nd ed., 1951, Prentice-Hall.
I (1–19, 47–51) ; II (21–28, 235) ; III (7–11, 74–77) ; IV (100–109) ; V (66–71) ; VI (155–166) ; VII (71–74) ; VIII (167–179) ; IX (145–150) ; X (none) ; XI (150–154) ; XII (28–30) ; XIII (30–37, 51–53) ; XIV (32–37, 180–185) ; XV (30–45, 180–199) ; XVI (54–60) ; XVII (40–45, 53–61) ; XVIII (21–28, 137–144) ; XIX (79–98) ; XX (109–128) ; XXI (130–135) ; XXII (212–221) ; XXIII (215–221) ; XXIV (221–225) ; XXV (225–230).

Elliott, Miles, Reynolds, *Mathematics: Advanced Course*, 1962, Prentice-Hall.
I (1–20) ; II (21–46) ; III (79–98) ; IV (100–109) ; V (109–136) ; VI (155–166) ; VII (212–225) ; VIII (167–179) ; IX (28–30) ; X (30–46, 180–200) ; XI (51–63) ; XII (40–45) ; XIII (137–144) ; XIV (201–212, 225–232) ; XV (145–154).

Evans, *Fundamentals of Mathematics*, 1959, Prentice-Hall.
I–VII (1–19) ; VIII (167–178) ; IX (1–19) ; XI (1–19) ; XII (1–19) ; XIII (47–51, 90–98, 178) ; XIV (100–112).

Freund, *A Modern Introduction to Mathematics*, 1956, Prentice-Hall.
I (Preface) ; II (1) ; III (1–5) ; IV (none) ; V (1–5) ; VI (1–3) ; VII (1–3) ; VIII (1–3, 237–240) ; IX (1–2, 13–18) ; X (1–3, 13–18) ; XI (147–148) ; XII (145–148) ; XIII (47–51, 89–93) ; XIV (100–104) ; XV (232–237) ; XVI (none) ; XVII (23–28, 79–87) ; XVIII (21–46) ; XIX (28–46) ; XX (212–231) ; XXI (none) ; XXII (none) ; XXIII (none) ; XXIV (153–154).

Helton, *Introducing Mathematics*, 1958, Wiley.
I (Preface) ; II (1–3) ; III (1–12) ; IV (13–18) ; V (1–5) ; VI (1–6) ; VII (1–7) ; VIII (none) ; IX (1–19) ; X (1–19) ; XI (1–19) ; XII (47–51) ; XIII (1–19) ; XIV (1–19) ; XV (1–19) ; XVI (100–104) ; XVII (21–28, 79–94) ; Part III (none).

Hill, Linker, *Introduction to College Mathematics*, rev. ed., 1955, Harper.
I (1–19, 23–30, 137–144) ; II (21–45) ; III (24–28, 157–158, 38–40) ; IV (12–16) ; V (7–11, 16–19, 66–71) ; VI (167–179) ; VII (180–185) ; VIII (49–51, 89–98) ; IX (100–109) ; X (155–166) ; XI (51–61); XII (185–199) ; XIII (74–77) ; XIV (150–154) ; XV (none) ; XVI (79–89) ; XVII (109–113) ; XVIII (115–135) ; XIX (137–144) ; XX (212–225) ; XXI (225–230).

Jaeger, Bacon, *Introductory College Mathematics*, 2nd ed., 1962, Harper.
I (21–23, 232–237) ; II (21–25) ; III (64–66) ; IV (79–99) ; V (100–114); VI (155–166) ; VII (66–78, 145–154) ; VIII (21–25) ; IX (212–225); X (225–231) ; XI (167–179) ; XII (25–46, 180–200) ; XIII (47–63) ; XIV (212–231) ; XV (212–231) ; XVI (74–79) ; XVII (24–28, 79–93, 115–136) ; XVIII (137–144) ; XIX (201–211).

Leonhardy, *Introductory College Mathematics*, 1963, Wiley.
I (232–235) ; II (1–19) ; III (7–15, 237–240) ; IV (167–179) ; V (3, 64–65) ; VI (21–27, 47–51, 235, 79–93, 100–104, 115–136, 155–166); VII (65–66) ; VIII (21–26, 212–230) ; IX (167–179) ; X (28–61, 180–200) ; XI (none) ; XII (146–154).

Lister, Rio, Sanders, *Freshman Mathematics for University Students*, 1966, Prentice-Hall.
I (1–19) ; II (232–235) ; III (1–19) ; IV (47–51, 71–74) ; V (21–28; 79–93, 100–136) ; VI (28–45, 51–62, 137–143, 180–200) ; VII (21–27, 47–50, 167–179) ; VIII (none) ; IX (64–72).

Mainardi, Konove, Baker, *A First Course in Mathematics*, 3rd ed., 1950, Van Nostrand.
I (1–4) ; II (28–53) ; III (180–185) ; IV (196–199) ; V (1–24) ; VI (49–51, 79–85, 65–66) ; VII (89–96) ; VIII (7–19) ; IX (167–179) ; X (28–61) ; XI (180–199) ; XII (100–104) ; XIII (74–77) ; XIV (155–156, 40–45, 53–61) ; XV (145–150) ; XVI (221–225) ; XVII (21–28, 79–87) ; XVIII (79–89) ; XIX (24–27) ; XX (109–128) ; XXI (130–135) ; XXII (137–144) ; XXIII (201–209) ; XXIV (212–230).

Meserve, Sobel, *Introduction to Mathematics*, 1964, Prentice-Hall.
I (245–248) ; II (1–19, 237–240) ; III (1–19) ; IV (232–235) ; V (23–29) ; 180–211); VII (1–19, 71–74) ; VIII (47–114) ; IX (none) ; X (23–30, 79–89, 109–144).

Milne, Davis, *Introductory College Mathematics*, 3rd ed., 1962, Ginn.
I (1–20) ; II (21–46) ; III (212–225) ; IV (225–231) ; V (79–99) ; VI (100–114) ; VII (155–166) ; VIII (214–225) ; IX (167–179) ; X (23–63) ; XI (180–200) ; XII (212–231) ; XIII (137–144) ; XIV (none) ; XV (167–170, 212–231) ; XVI (none) ; XVII (79–99, 241–242) ; XVIII (115–136); XIX (130–136) ; XX (201–212) ; XXI (none); XXII (145–154) ; XXIII (145–150) ; XXIV (232–245).

Moore, *Fundamental Principles of Mathematics*, 1960, Rinehart.
I (232–237) ; II (1–5) ; III (167–179, 237–240) ; IV (21–30) ; V (30–46) ; VI (79–89) ; VII (27, 51–63, 180–200) ; VIII (212–225) ; IX (145–150, 225–231) ; X (74–78, 155–166) ; XI (90–99, 104–109) ; XII (104–136) ;

XIII (137–144); XIV (none); XV (201–211); XVI (150–154, 66–71); XVII (153–154).

Newsom, Eves, *Introduction to College Mathematics*, 2nd ed., 1954, Prentice-Hall.

I (1–18); II (7–11); III (none); IV (167–179); V (145–150); VI (none); VII (150–154); VIII (21–27, 235); IX (65–66); X (28–45); XI (28–35, 137–139, 180–199); XII (49–51, 100–104, 155–166); XIII (79–89, 109–113, 115–135); XIV (212–230).

Rees, *Freshman Mathematics*, 1959, Prentice-Hall.

I (1–5); II (21–28); III (47–51, 79–98); IV (28–46); V (79–98); VI (5–15); VII (15–19); VIII (180–185); IX (7–12); X (100–113); XI (167–178); XII (180–200); XIII (none); XIV (145–150); XV (64–66); XVI (none); XVII (150–154).

Richardson, *Fundamentals of Mathematics*, 3rd ed., 1965, Macmillan.

I (v–vi); II (232–245); III (1–7); IV (7–19, 71–77); V (7–18, 47–51, 79–113); VI (1–5, 237–241, 167–179); VII (145–154); VIII (none); IX (23–28, 79–91, 100–144, 201–211); X (21–27, 79–91, 232–237); XI (212–230); XII (28–46, 51–62, 180–200); XIII (none); XIV (none); XV (66–71); XVI–XIX (none).

Rider, *First Year Mathematics for Colleges*, 2nd ed., 1962, Macmillan.

I (1–20); II (12–13); III (13–17); IV (21–25); V (21–46); VI (47–50, 79–99); VII (7–10); VIII (100–114); IX (104–114); X (71–74); XI (64–66); XII (66–72); XIII (145–150); XIV (1–3, 237–244); XV (167–179); XVI (none); XVII (none); XVIII (none); XIX (24–40); XX (180–200); XXI (180–200); XXII (25–46); XXIII (180–200); XXIV (47–63); XXV (28–29); XXVI (29–46); XXVII (40–46); XXVIII (47–63); XXIX (18, 74–78); XXX (155–166); XXXI (155–166); XXXII (47–55); XXXIII (79–99); XXXIV (109–114); XXXV (115–136); XXXVI (222); XXXVII (none); XXXVIII (137–144); XXXIX (none); XXXX (147–154); XXXXI (150–154); XXXXII (94–99, 241–243); XXXXIII (none); XXXXIV (147–154); XXXXV (201–212); XXXXVI (201–212); XXXXVII (201–212).

Rose, *A Modern Introduction to College Mathematics*, 1959, Wiley.

I (232–237); II (21–23, 232–237); III (1–5, 27); IV (1–5); V (47–51, 71–77, 100–104); VI (21–28, 79–98); VII (28–45); VIII (51–61, 180–199); IX (108–113); X (109–136); XI (none); XII (none); XIII (137–144); XIV (212–230); XV (none).

Sachs, Rasmusen, Purcell, *Basic College Mathematics*, 2nd ed., 1965, Allyn and Bacon.

I (232–236); II–V (1–20, 71–73); VI (21–28, 79–99); VII (7–9, 100–102); VIII (100–109); IX (27, 64–66); X (150–154); XI (221–223).

Smith, Schroeder, *College Mathematics for Freshmen*, 1959, Van Nostrand.

I (1–5); II (4–5); III (none); IV (1–19); V (1–19); VI (47–51); VII (70–71); VIII (47–51, 89–98); IX (21–45); X (7–12); XI (100–109); XII (167–178); XIII (none); XIV (none); XV (64–66); XVI (180–184); XVII (185–199); XVIII (none).

Taylor, Wade, *University Freshman Mathematics,* **1963, Wiley.**
I (232–237, 21); II (1–26); III (1–20); IV (2, 21–28, 235–237); V (79–98); VI (96–98, 241–243, 196, 74–76); VII (66–71); VIII (47–50, 100–104, 155–166); IX (7–11, 167–179); X (28–45, 51–63); XI (40–46); XII (74–78); XIII (180–200).

Trimble, Lott, *Elementary Analysis,* **1960, Prentice-Hall.**
I (1–5, 74–76); II (1–5); III (232–237); IV (21–46, 79–86, 145–150); V (7–10, 74–78); VI (47–51, 100–104); VII (89–99); VIII (167–179); IX (28–46, 51–63, 180–200); X (79–89, 113–144, 201–211); XI (none).

Vance, *Fundamentals of Mathematics,* **1960, Addison-Wesley.**
I–II (1–20); III–IV (21–63); V (79–89); VI (100–113); VII (89–100, 241–242); VIII (69–73); IX–X (212–221); XI (155–166); XII (221–224); XIII (40–45); XIV (167–179); XV (180–199); XVI (40–45); XVII (74–78); XVIII (115–136); XIX (223–231); XX (150–154).

Wade, *Introductory College Mathematics,* **1959, Wiley.**
I (1–19); II (1–19); III (21–28, 79–85, 109–113); IV (79–98); V (115–135); VI (137–144); VII (21–31, 232–237); VIII (212–225).

Wade, Taylor, *Fundamental Mathematics,* **1961, McGraw-Hill.**
I (1–5); II (1–10); III (1–17); IV (47–51); V (64–66); VI (none); VII (21–38); VIII (7–10); IX (64–66); X (100–108); XI (21–30, 79–92); XII (none); XIII (none); XIV (167–177); XV (145–150); XVI (66–78, 100–108).

Western, Haag, *An Introduction to Mathematics,* **1959, Holt.**
I (232–237); II (1–19); III (1–19); IV (1–19, 47–51, 100–104, 155–160); V (167–178); VI (232–237); VII (21–28); VIII (none); IX (28–45, 74–77, 137–144); X (79–94); XI (115–135); XII (212–230); XIII (212–230); XIV (212–230); XV (221–223); XVI (none); XVII (201–211).

Chapter in This Outline	TOPIC	Allendoerfer & Oakley *Fund.*	Allendoerfer & Oakley *Prin.*	Apostle	Ayres, Fry, Jonah	Banks
I	Fundamental Concepts	37, 38 94, 106	48	69, 83 99, 107	140, 239	47, 117
II	Functions and Their Representation	212, 240 263, 284	124, 187 232	274, 325	26, 83 118	179, 255 416
III	Equations and Identities	121, 320	23, 136	117, 367	44	174
IV	Algebraic Tools	51, 82 187	87, 435		11	37, 222 334
V	The Linear Function	126, 378	136, 276	117, 377	32	352, 392
VI	Quadratic Equations	126	152	149	59	400
VII	The Conic Sections	384	296	384	69, 72	361
VIII	Polar Coordinates	398	309			
IX	Progressions		402	253, 352		
X	Theory of Equations	257	162			
XI	Logarithms	264	223	310	148	299
XII	Solutions of Triangles	305		329, 333	74, 89 105	262
XIII	Three-dimensional Space			392		
XIV	The Calculus	412, 446	320, 367	396		
XV	Additional Topics	1	8, 447	3	264, 288	232

TO STANDARD TEXTBOOKS

refer to pages.

Brink	Brixey & Andree	Denbrow & Goedicke	Elliott & Miles	Elliot, Miles, Reynolds	Evans	Freund
1, 598	1, 173	47	3, 30	1	32, 55 80	7, 16
21, 30 193	15, 322	89, 338	19, 165 172, 194	38, 235 243, 303		256, 289 306
108, 248 259	20, 356	166, 309 361	12, 212 224	28, 282	232	195, 502
68, 101 155, 314	42, 231 257		58, 89	47, 374		493
40, 46 357	103	397	252	72	237	195
77, 122 397	23, 482	227 391	42	92	256	210
384, 419	490	415	270	122		281
304	394, 524		244	311		
134, 453 467, 515	215, 248	221, 273	122	369		183
328	79		64	154		
169	182	200	107	216	155	487
274	335		188, 203	264		512
564	560			349		
	130, 429	442	309, 328 343, 362	181, 337		334
497	1, 583					400, 424

All figures

Chapter in This Outline	TOPIC	Helton	Hill & Linker	Jaeger & Bacon	Leon-hardy	Lister, Rio, & Sanders
I	Fundamental Concepts	12, 154	1, 69 80	94	37, 99	2, 68 94
II	Functions and Their Representation	262	7, 16 53	6, 12	184, 333	116, 165 225
III	Equations and Identities	191	7, 169	228, 338	63, 341	184
IV	Algebraic Tools		88, 199	31, 117	165, 244	348
V	The Linear Function		112, 267	42	200	131
VI	Quadratic Equations	251	132, 296	70	220	
VII	The Conic Sections		309	351	230	148
VIII	Polar Coordinates		331	387		165
IX	Progressions		214	121		
X	Theory of Equations		153	94	214	
XI	Logarithms		93	206	141, 311	289
XII	Solutions of Triangles		107, 181	233	371	193, 219
XIII	Three-dimensional Space	362		406		
XIV	The Calculus		341, 382	153, 180	260	
XV	Additional Topics			1	4, 89	4, 46

TO STANDARD TEXTBOOKS *(cont.)*

refer to pages.

Mainardi, Konove, Baker	Meserve & Sobel	Milne & Davis	Moore	Newsom & Eves	Rees
99	23, 127	1	27, 45	1, 25	1
17, 58 223	78, 172	33, 183	64, 126 351	164, 202	26, 58
131, 184	156	183	174	254	40, 58
170		420	470	187	206
71, 86 236	165, 223	101, 324	317	284	40, 71
156		116	277, 342	270	134
266, 287	235	347	370	300	
303		261	380	233	
198	101	413		91, 146	194, 227
185		132	277	277	
111		170, 283	53	65	150
36, 140		214	195	236	107, 164
319	89	386	442		
209, 337		51, 84	211, 255	322	
	1, 44 62		1		

Chapter in This Outline	TOPIC	Richard-son	Rider	Rose	Sachs *et al.*	Smith & Schroeder
I	Fundamental Concepts	49, 129 157	1, 79	70	35, 71 131, 189	1, 18 36, 51
II	Functions and Their Rep-resentation	255, 307	35, 49 202, 225	18, 137 187	161	107, 190
III	Equations and Identities	383	259, 355	105	161	64
IV	Algebraic Tools	499	137, 142		218	86, 184
V	The Linear Function	136, 150 281	55, 515	137	161	70, 88
VI	Quadratic Equations	142, 284	97	129, 300	235	133
VII	The Conic Sections		396	300	241	
VIII	Polar Coordinates		460	376		
IX	Progressions	222, 419	152, 497		332	172
X	Theory of Equations		332			
XI	Logarithms	177	169	346		147
XII	Solutions of Triangles	390	210, 235			190, 200
XIII	Three-dimen-sional Space	302	550, 559			212
XIV	The Calculus	331		389		
XV	Additional Topics	6		3		

TO STANDARD TEXTBOOKS *(cont.)*

refer to pages.

Taylor & Wade	Trimble & Lott	Vance	Wade	Wade & Taylor	Western & Haag
13, 220	1, 43	1, 21	1, 64	1, 21	33
7	73, 95 408	36	98, 227	183, 253	173, 246
195, 286	220	60	300	115, 394	101, 246
174		146, 169	91	211	56, 114
122	95, 220 320	99	132		293
	229	126, 334	164	224	
	301	343	164		328
	461, 534		213		
	113, 144	270, 399		337	
200					112
220	370	269		319	144
323	419	297			
151	586				476
		180, 203 375	268		343, 371
1, 156	73				1

GREEK ALPHABET

Letters		Names	Letters		Names	Letters		Names
A	α	Alpha	I	ι	Iota	P	ρ	Rho
B	β	Beta	K	κ	Kappa	Σ	σ	Sigma
Γ	γ	Gamma	Λ	λ	Lambda	T	τ	Tau
Δ	δ	Delta	M	μ	Mu	Υ	υ	Upsilon
E	ε	Epsilon	N	ν	Nu	Φ	φ	Phi
Z	ζ	Zeta	Ξ	ξ	Xi	X	χ	Chi
H	η	Eta	O	o	Omicron	Ψ	ψ	Psi
Θ	θ	Theta	Π	π	Pi	Ω	ω	Omega

SYMBOLS

$=$, is equal to;

$\neq$, is not equal to;

$<$, is less than;

$>$, is greater than;

$\leq$, is less than or equal to;

$\geq$, is greater than or equal to;

ln, natural logarithm;

log, common logarithm;

$\frac{d}{dx}(\ )$, derivative of;

P_1, P sub 1;

$n \to \infty$, n approaches infinity;

$\sqrt{n}$, square root of n;

$\angle ABC$, angle with vertex at B;

$\triangle ABC$, triangle ABC;

$\cong$, is congruent to;

Δx, increment of x;

$f(x)$, function of x;

$\int$, integral of.

1

FUNDAMENTAL CONCEPTS

1. Introduction. We are about to study the elementary phase of college mathematics. In order to do this, it is necessary to understand thoroughly the fundamentals upon which the study is based. Many of these concepts are already familiar to the student, and consequently the first chapter will be in the nature of a review. However, the student should not take this chapter lightly, as its contents will be in continual use, and the key to understanding mathematics is to be very proficient in the use of the fundamental concepts. It is strongly recommended that the student have a complete understanding of this chapter before proceeding further.

The importance of mathematics in the modern world is so well established that the student should not need any further inspiration to want to know as much as possible about this subject. But in case the reader is having some trouble, he is urged to develop an enthusiasm within himself. A spirited endeavor will make the work easier and a successful completion of the study far more rewarding.

2. The Real Numbers. The first thing to consider in mathematics is the number system. Thus, in arithmetic we meet the integers, fractions, and decimals. We should be familiar with:

the *positive integers*	$1, 2, 3, 4, 5, \ldots$
the *zero integer*	0
the *negative integers*	$-1, -2, -3, -4, -5, \ldots$
the *fractions*	$\frac{1}{2}, \frac{2}{3}, \frac{7}{8}, \ldots$
the *decimal fractions*	$.25, .01, .0024, \ldots$

Then we can define:

a **rational number** *as one which can be expressed as the quotient of two integers*, such as 5 (which is $\frac{5}{1}$ understood), $\frac{3}{4}$, $-\frac{2}{3}$, etc.

Furthermore, we define:

an **irrational number** *as one which cannot be expressed exactly as the quotient of two integers*. There are many such irrational numbers, for

1

example, $\sqrt{2}$, $\sqrt[3]{3}$, π, etc. These are usually expressed as unending or nonrepeating decimals, $\sqrt{2} = 1.414\ldots$, or approximated correctly to a certain number of decimal places, $\pi = 3.1416$.

In forming quotients for the real numbers we exclude the division by zero since it has no meaning. Thus we have now defined the *real-number system.*

To complete our discussion of real numbers for the present we define the **absolute value** of a number as *its numerical value without regard to sign;* symbolically, the absolute value is represented by two vertical lines on each side of the number. Thus we say the absolute value of -3 is equal to 3 or $|-3| = 3$. Another example is $|5| = 5$. Thus $|a| = a$ if a is positive or zero, and $|a| = -a$ if a is negative.

Later we shall define an imaginary number, which will lead us to complex numbers.

The arithmetic of numbers should be familiar to the student. We shall briefly summarize a few operations which seem to be troublesome.

1. To add or subtract two or more fractions, find the lowest common denominator (L.C.D.), change each fraction to its equivalent over the L.C.D., and add the numerators.

Illustration.

$$\frac{2}{3} + \frac{3}{4} - \frac{5}{6} = \frac{2(4) + 3(3) - 5(2)}{12} = \frac{8 + 9 - 10}{12} = \frac{7}{12}.$$

2. To multiply two or more fractions, multiply numerator by numerator and denominator by denominator; reduce the result to its lowest form.

Illustration.

$$\frac{2}{3} \times \frac{3}{4} \times \frac{5}{6} = \frac{2 \times 3 \times 5}{3 \times 4 \times 6} = \frac{30}{72} = \frac{5}{12}$$

or

$$\frac{2}{3} \times \frac{3}{4} \times \frac{5}{6} = \frac{\overset{1}{\cancel{2}} \times \overset{1}{\cancel{3}} \times 5}{\underset{1}{\cancel{3}} \times \underset{2}{\cancel{4}} \times 6} = \frac{1 \times 1 \times 5}{1 \times 2 \times 6} = \frac{5}{12}.$$

3. To divide two fractions, multiply the numerator by the reciprocal of the denominator.

Illustration.

$$\frac{2}{3} \div \frac{5}{7} = \frac{2}{3} \times \frac{7}{5} = \frac{14}{15}.$$

4. To add numbers containing decimals, add the corresponding digits, retaining the position of the decimal point.

Illustration. Find the sum of 1.23 + 6.5 + 0.149.
Solution.

$$\begin{array}{r} 1.230 \\ 6.500 \\ 0.149 \\ \hline 7.879 \end{array}$$

5. To multiply numbers containing decimals, multiply each digit of the multiplicand by each digit of the multiplier and add the number of decimal places to obtain the position of the decimal point in the product.

Illustration. Multiply 3.14 by 2.16.
Solution.

$$\begin{array}{r} 3.14 \quad \text{(multiplicand)} \\ 2.16 \quad \text{(multiplier)} \\ \hline 1\,8\,8\,4 \\ 3\,1\,4 \\ 6\,2\,8 \\ \hline 6.7\,8\,2\,4 = 6.78 \quad \text{(product)} \end{array}$$

6. To obtain a given per cent of a number, change the per cent to its decimal equivalent and multiply.

Illustration. What is $2\frac{1}{2}\%$ of 325.00?
Solution. $2\frac{1}{2}\% = .025.$

$$N = (.025)(325.00) = 8.125.$$

The student should "brush-up" on his multiplication tables, the powers of the integers, and the roots of the integers. Table I in the back of this book will be useful, but the advantage of memorizing the following table will also become apparent throughout the study of this book.

N \ Power	2	3	4	5	6	7	8	9
2	4	8	16	32	64	128	256	512
3	9	27	81	243	729			
4	16	64	256	1024				
5	25	125	625					
6	36	216						

3. Representation of Numbers. In order to obtain a better understanding of the real-number system, let us picture it graphically. Consider a straight line of unlimited length and choose a point 0, which we shall call the **origin.** Let the direction be chosen *positive* if taken to the right and *negative* if taken to the left of this point. Then choose an arbitrary length as the unit of measure and lay off successive units in both directions from the origin. If these now be labeled 1, 2, 3, ... and −1, −2, −3, ... we have constructed a number scale as shown in Fig. 1.

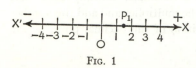

FIG. 1

On this number scale, one can locate any real number whether it be rational or irrational. The rational numbers are easily located since for them it is simply a matter of dividing up the unit length. As an example, the point P_1 represents $\frac{3}{2}$ and was obtained by halving the unit length between 1 and 2. The student has had many experiences with the location of rational numbers in his everyday handling of rulers. The location of an irrational number may not be so simple, but it can be done. Thus the number $\sqrt{2}$ could be located in its decimal form to the accuracy of being able to measure the distance 1.414, or it could be located exactly by a geometric construction. We are led to this very important statement:

To every real number there corresponds one and only one point on a line X′X, and, conversely, to every point on the line X′X there corresponds one and only one real number.

It is not easy to prove this statement, but the student should comprehend its meaning. We are now in a position to visualize graphically the meaning of one number being larger or smaller than another, for a number *larger* than another is always to the *right* of it and a number *smaller* is always to the *left* of it. This may be tested by comparing the positions of the numbers 2 and 3 or −1 and −3 on the line of Fig. 1.

4. The Formula. It may be remembered that the area of a rectangle is obtained by multiplying the length by the width. In the notation of mathematics this fact may be written $A = ab$, where A represents the area, a represents the length, and b represents the width. This is a simple introduction to the algebraic formula, in which letters are used to represent numbers and specify a well-known fact. The letters have different values depending upon the problem under discussion, and their meaning must be clearly understood. They are frequently spoken of as *general numbers*. The study of algebra is essentially concerned with the operations of, establishing laws for, and understanding the meaning of such general numbers.

Illustrations.

$$\text{Area of a circle:} \quad A = \pi r^2.$$
$$\text{Area of a triangle:} \quad A = \tfrac{1}{2}bh.$$

Furthermore, we shall be concerned with combinations of symbols. If each symbol in a combination represents a number, we shall call the combination an **expression.**

Illustrations.

$$2ax, \quad 2x - 3a, \quad 4y^2 + 3ax - 12.$$

If there are several parts connected by plus and minus signs, each part is called a **term.** Thus, in the above illustrations, $2x$ is a term and $-3a$ is a term. Note that the sign is associated with the term, but if the sign is $+$ it is usually not expressed unless for emphasis.

In discussing a term of an expression we have additional terminology. Thus we say that each quantity of two or more that are multiplied together is a **factor** of the product. Furthermore, any factor of an expression is called the **coefficient** of the remaining part.

Illustration.

$$32bx^2$$

32, b, and x^2 are factors.
32 is the numerical coefficient.
$32b$ is the coefficient of x^2.
$32x^2$ is the coefficient of b.

Names are also given to various expressions. Thus we have:

    monomial, an expression consisting of one term,
    binomial, an expression consisting of two terms,
    trinomial, an expression consisting of three terms,
    polynomial, an expression consisting of more than one term.

5. Fundamental Operations. It is necessary that we establish some ground rules in order that we may proceed in an orderly manner. We start with the four fundamental operations of addition, subtraction, multiplication, and division. First we have three important laws:

I. The commutative law. *The result of addition or multiplication is the same in whatever order the terms are added or multiplied.*

$$a + b = b + a, \quad \text{and} \quad ab = ba.$$

II. The associative law. *The sum of three or more terms, or the product of three or more factors, is the same in whatever manner they are grouped.*

$$a + (b + c) = (a + b) + c = a + b + c.$$
$$a(bc) = (ab)c = abc.$$

III. The distributive law. *The product of an expression of two or more terms by a single factor is equal to the sum of the products of each term of the expression by the single factor.*

$$(a + b - c)d = ad + bd - cd.$$

Next, we have some **laws of signs:**

I. To add two numbers having like signs, *add their absolute values and prefix the common sign.*

Illustrations.

$$3 + 5 = 8; \quad -3 + (-5) = -8.$$

II. To add two numbers having unlike signs, *take the difference of their absolute values and prefix to it the sign of the number having the larger absolute value.*

Illustrations.

$$5 + (-2) = 3; \quad 2 + (-5) = -3.$$

III. To subtract one number from another, *change the sign of the number to be subtracted and proceed as in addition.*

Illustrations.

$$7 - (+3) = 7 + (-3) = 4;$$
$$3 - (+7) = 3 + (-7) = -4;$$
$$3 - (-7) = 3 + (+7) = 10.$$

IV. The product, or quotient, of two numbers *has the following rule of signs.*

a positive *by a* positive *yields a* positive;
a positive *by a* negative *yields a* negative;
a negative *by a* positive *yields a* negative;
a negative *by a* negative *yields a* positive.

Illustrations.

$$3 \times 2 = 6; \quad 6 \div (-2) = -3;$$
$$-4 \times 2 = -8; \quad (-8) \div (-4) = 2.$$

In dealing with algebraic expressions it is often necessary to group together several parts. This is accomplished by using **parentheses,**

( ); **brackets,** [ ]; **braces,** { }, and the **vinculum,** —. The last one is not recommended. These symbols of grouping may be removed or inserted if the following rules are obeyed:

Symbols of grouping preceded by a plus sign *may be removed by rewriting each of the inclosed terms with its original sign.*

Illustration.

$$3x + (2y - 4z) = 3x + 2y - 4z.$$

Symbols of grouping preceded by a minus sign *may be removed by rewriting and changing the sign of each of the inclosed terms.*

Illustration.

$$3x - (2y - 4z) = 3x - 2y + 4z.$$

If a **coefficient** *precedes a symbol of grouping, it is to be multiplied into each included term when the symbol is removed.*

Illustration.

$$x - 2(3y - 2a) = x - 6y + 4a.$$

To simplify expressions involving several symbols of grouping we work from the inside out by *first removing the innermost pair* of symbols, next the innermost pair of the remaining ones, and so on. If like terms appear they are combined at each step as in the following example.

Illustration.

$$4a - b - \{3a - [2a(4 - b) - (a - b)]\}$$
$$= 4a - b - \{3a - [8a - 2ab - a + b]\}$$
$$= 4a - b - \{3a - 7a + 2ab - b\}$$
$$= 4a - b + 4a - 2ab + b$$
$$= 8a - 2ab.$$

6. Powers, Exponents, and Radicals. In the case of multiplying together factors which are all alike, a shorthand method of writing the product has been devised, and the product is called a **power** of the factor. Thus, $x \cdot x \cdot x$ is the *third* power of x and we write it as x^3. The number x is called the **base** and 3 the **exponent** of the power. Thus the exponent is the number of times the like factor appears in the multiplication. We speak of:

$$x^2 \quad \text{as} \quad x \text{ square,}$$
$$x^3 \quad \text{as} \quad x \text{ cube,}$$
$$x^n \quad \text{as} \quad \text{the } n^{\text{th}} \text{ power of } x.$$

Illustrations.

$$(-2)(-2)(-2)(-2) = (-2)^4 = 16.$$
$$x^5 = x \cdot x \cdot x \cdot x \cdot x = \text{five factors of } x.$$

In the above discussion we have limited ourselves to the case where the exponent is a positive integer. We shall generalize this, but before we do so, let us consider the **basic laws of exponents.** In the following let m and n be any two positive integers.

> I. $a^m \cdot a^n = a^{m+n}$.
>
> II. $(a^m)^n = a^{mn}$.
>
> III. $(ab)^n = a^n b^n$.
>
> IV. $\left(\dfrac{a}{b}\right)^n = \dfrac{a^n}{b^n}, \quad b \neq 0.$
>
> V. $\dfrac{a^m}{a^n} = a^{m-n}, \quad m > n, \quad a \neq 0.$
>
> VI. $\dfrac{a^m}{a^n} = \dfrac{1}{a^{n-m}}, \quad m < n, \quad a \neq 0.$

These laws may be proved by considering the definitions, e.g.:

$$a^m \cdot a^n = \underbrace{(a \cdot a \cdot a \cdots a)}_{m \text{ factors}}\underbrace{(a \cdot a \cdot a \cdots a)}_{n \text{ factors}}$$
$$= (a \cdot a \cdot a \cdots a), \quad m + n \text{ factors}$$
$$= a^{m+n}.$$

The student may like to prove the others by writing out the definition and collecting terms.

Illustrations.

$$2^3 \cdot 2^5 = 2^8 = 256.$$
$$(3^2)^3 = 3^6 = 729.$$
$$(2 \cdot 3)^2 = 2^2 \cdot 3^2 = 4 \cdot 9 = 36.$$
$$\left(\frac{2}{3}\right)^2 = \frac{2^2}{3^2} = \frac{4}{9}.$$
$$\frac{2^5}{2^2} = 2^3 = 8.$$
$$\frac{3^2}{3^3} = \frac{1}{3}.$$

The next logical step is to define a root, which we do by the statement that if $a^m = N$, then a is the mth **root of** N. That is, we are looking for a number a such that when multiplied by itself to m factors it yields N. We write this statement with the use of a **radical** sign, $\sqrt{}$, thus:

$$\sqrt[m]{N} = a.$$

The process of taking a root does not necessarily lead to a single answer. For example, the square root of 4 may be either $+2$ or -2.

Illustrations.

$$\sqrt{16} = \pm 4; \quad \sqrt[4]{81} = \pm 3.$$
$$\sqrt[3]{-8} = -2; \quad \sqrt[5]{32} = 2.$$

In the symbol $\sqrt[q]{N}$, q is called the **index of the root** and N, the **radicand**. The radicals obey the same laws as exponents, but before we turn to the proof we must first consider exponents which are not integers. We begin with the *fractional exponents*. Let $m = 1/q$, then according to Law I we may write:

$$\underbrace{a^{\frac{1}{q}} \cdot a^{\frac{1}{q}} \ldots a^{\frac{1}{q}}}_{q \text{ factors}} = a^{\frac{q}{q}} = a,$$

which is our definition of the qth root of a, and we have:

$$(a^{\frac{1}{q}} = \sqrt[q]{a}.)$$

If we raise this expression to the pth power, we write:

$$\left(a^{\frac{1}{q}}\right)^p = a^{\frac{p}{q}} = (\sqrt[q]{a})^p,$$

and we have the complete definition of a fractional exponent, which conforms to the laws of exponents.

Illustrations.

$$4^{\frac{1}{2}} = \sqrt{4} = 2; \quad 27^{\frac{1}{3}} = 3.$$

If we let $m = 0$, we have, according to Law I:

$$a^0 \cdot a^n = a^{0+n} = a^n$$

or

$$a^0 = \frac{a^n}{a^n} = 1, \quad a \neq 0,$$

so that the definition is:

$$a^0 = 1, \quad a \neq 0.$$

Illustrations.

$$(3x)^0 = 1, \quad (x \neq 0); \quad 5y^0 = 5, \quad (y \neq 0).$$

If we let $m = -n$, where n is positive, we may write, according to Law I:

$$a^{-n} \cdot a^n = a^{-n+n} = a^0 = 1, \quad a \neq 0,$$

so that the definition reads:

$$a^{-n} = \frac{1}{a^n}, \quad a \neq 0.$$

Illustrations.

$$2x^2 y^{-3} = \frac{2x^2}{y^3}; \quad 3(xy)^{-2} = \frac{3}{x^2 y^2}.$$

We have now defined exponents and roots for all practical purposes. The student should write down illustrations and have a thorough knowledge of these definitions before reading further. It should also be noted that the laws of exponents apply to radicals since radicals are simply fractional exponents. In particular, the following four laws may be emphasized (a and b are positive):

I. $(\sqrt[n]{a})^n = a.$

II. $(\sqrt[n]{a})(\sqrt[n]{b}) = \sqrt[n]{ab}.$

III. $\sqrt[m]{\sqrt[n]{a}} = \sqrt[mn]{a}.$

IV. $\dfrac{\sqrt[n]{a}}{\sqrt[n]{b}} = \sqrt[n]{\dfrac{a}{b}}.$

The simplification of radicals usually reduces to one of the following operations:

1. **Removing factors from the radicand.**

Illustration.

$$\sqrt{50x^3} = \sqrt{(25x^2)2x} = 5x\sqrt{2x}.$$

2. **Introducing coefficients under the radical sign.**

Illustration.

$$2a\sqrt[3]{3b} = \sqrt[3]{(2a)^3}\sqrt[3]{3b} = \sqrt[3]{(8a^3)(3b)} = \sqrt[3]{24a^3b}.$$

3. **Reducing to a radical of lower order.**

Illustration.

$$\sqrt[6]{81} = \sqrt[6]{3^4} = 3^{\frac{4}{6}} = 3^{\frac{2}{3}} = \sqrt[3]{3^2} = \sqrt[3]{9}.$$

4. **Rationalizing the denominator.**

Illustrations.

$$\sqrt{\frac{3}{5}} = \sqrt{\frac{3}{5} \cdot \frac{5}{5}} = \frac{\sqrt{15}}{\sqrt{25}} = \frac{1}{5}\sqrt{15};$$

$$\frac{\sqrt{2}}{\sqrt{7}} = \frac{\sqrt{2}}{\sqrt{7}} \cdot \frac{\sqrt{7}}{\sqrt{7}} = \frac{1}{7} \sqrt{14}.$$

The above four operations on radicals will be used again in Section 11.

7. Multiplication and Division. The multiplication of algebraic expressions is easily accomplished by paying strict attention to the laws of operations discussed in Section 5 and the laws governing exponents discussed in Section 6. Let us begin by operating on polynomials, which are expressions consisting of more than one term each. Then we have:

To find the **product of two polynomials,** *multiply each term of the* **multiplicand** *by each term of the* **multiplier** *and add the results.*

It is important to arrange the work systematically, and it is suggested that the multiplicand and multiplier be arranged in descending powers of one letter.

Illustration.   Multiply $(3x^2 - y^2 + 2xy)$ by $(y^2 - x^2 - 3xy)$.
Solution.

$$
\begin{array}{llll}
3x^2 + & 2xy - & y^2 & \text{(multiplicand)} \\
-x^2 - & 3xy + & y^2 & \text{(multiplier)} \\
\hline
-3x^4 - & 2x^3y + & x^2y^2 & \\
 & -9x^3y - & 6x^2y^2 + 3xy^3 & \\
 & & 3x^2y^2 + 2xy^3 - y^4 & \\
\hline
-3x^4 - & 11x^3y - & 2x^2y^2 + 5xy^3 - y^4 & \text{(product)}
\end{array}
$$

To check a multiplication, replace the letters by numbers in the multiplicand, multiplier, and product.

Check.   Let $x = 2$ and $y = 3$ in above illustration.
Multiplicand $= 3(4) + 2(2)(3) - 9 = 15.$
Multiplier $= -4 - 3(2)(3) + 9 = -13.$
Product $= -3(16) - 11(8)(3) - 2(4)(9) + 5(2)(27) - 81 = -195.$
Multiplicand $\times$ multiplier $= (15)(-13) = -195.$

The operation of **division** is the inverse of multiplication. It is defined by the relation:

Dividend = divisor $\times$ quotient + remainder.

If the remainder is zero, the division is **exact** and the divisor is said to be a **factor** of the dividend. *To find the quotient of one polynomial divided by another* carry out the following steps:

1. Arrange each polynomial in descending powers of a common letter.

2. Divide the first term of the *dividend* by the first term of the *divisor*, yielding the first term of the *quotient*.

3. Multiply the *whole* divisor by the first term of the quotient, and subtract the product from the dividend.

4. The remainder is a new dividend, and the steps 2 and 3 are repeated on this dividend.

5. Continue the process until a remainder is obtained which is an expression whose first term does not contain the first term of the divisor as a factor.

6. Check the result by replacing the letters by numbers. Do not use numbers which make the divisor zero.

The work should be arranged schematically, and a scheme is suggested in the illustration.

Illustration. Divide $15x^3 + 23x - 26x^2 - 3$ by $3x^2 + 3 - 4x$.
Solution.

$$
\begin{array}{r|l}
\text{(dividend)}\quad 15x^3 - 26x^2 + 23x - 3 & 3x^2 - 4x + 3 \quad \text{(divisor)} \\
\underline{15x^3 - 20x^2 + 15x} & \overline{5x - 2} \qquad\quad \text{(quotient)} \\
- 6x^2 + 8x - 3 & \\
\underline{- 6x^2 + 8x - 6} & \\
3 \quad \text{(remainder)} &
\end{array}
$$

Check. Let $x = 2$.

Dividend $= 15(8) - 26(4) + 23(2) - 3 = 59$.
Divisor $= 3(4) - 4(2) + 3 = 7$.
Quotient $= 5(2) - 2 = 8$.
Remainder $= 3$.
Dividend $=$ divisor $\times$ quotient $+$ remainder.
$$59 = (7)(8) + 3 = 56 + 3 = 59.$$

8. Factoring. We shall now consider the process of finding two or more expressions whose product is a given expression. This process is called **factoring** and is in reality exact division or the inverse of multiplication. We accomplish this by being able to recognize a group of typical products, and upon seeing them in an algebraic expression we can immediately write down the factors. The first step, therefore, is to memorize the following products (the student may verify them by multiplying the right-hand factors):

1. $ab + ac = a(b + c)$.
2. $a^2 + 2ab + b^2 = (a + b)^2$.
3. $a^2 - 2ab + b^2 = (a - b)^2$.
4. $a^2 - b^2 = (a + b)(a - b)$.

5. $x^2 + (a + b)x + ab = (x + a)(x + b)$.
6. $a^3 + b^3 = (a + b)(a^2 - ab + b^2)$.
7. $a^3 - b^3 = (a - b)(a^2 + ab + b^2)$.
8. $acx^2 + (ad + bc)x + bd = (ax + b)(cx + d)$.
9. $a^2 + b^2 + c^2 + 2ab + 2ac + 2bc = (a + b + c)^2$.
10. $a^3 + 3a^2b + 3ab^2 + b^3 = (a + b)^3$.

With these basic products firmly in mind, it is possible to factor algebraic expressions by inspection. To ease the process of factoring, first remove all common expressions and then make advantageous grouping.

Illustrations.

(1) $\begin{aligned} 27x^2 + 36xy + 12y^2 &= 3(9x^2 + 12xy + 4y^2) \\ &= 3(3x + 2y)^2. \end{aligned}$

(2) $\begin{aligned} ax^2 - ay^2 - 6ay - 9a &= a(x^2 - y^2 - 6y - 9) \\ &= a[x^2 - (y^2 + 6y + 9)] \\ &= a[x^2 - (y + 3)^2] \\ &= a[x - (y + 3)][x + (y + 3)] \\ &= a(x - y - 3)(x + y + 3). \end{aligned}$

(3) $\begin{aligned} x^2y^2 - 4cxy - 2xy + 8c &= x^2y^2 - 2xy - 4cxy + 8c \\ &= xy(xy - 2) - 4c(xy - 2) \\ &= (xy - 2)(xy - 4c). \end{aligned}$

One of the uses of factoring is the finding of the lowest common multiple of a group of algebraic expressions. The **lowest common multiple (L.C.M.)** is defined as the expression which has the smallest number of factors contained in the given expressions. To find the L.C.M. we simply factor each expression and collect each factor the largest number of times it occurs in any one expression.

Illustration. Find the L.C.M. of $(3x - 6)$, $3x^2 - 12x + 12$, and $x^3 - 8$.
Solution.

$$3x - 6 = 3(x - 2).$$
$$3x^2 - 12x + 12 = 3(x - 2)(x - 2).$$
$$x^3 - 8 = (x - 2) \qquad (x^2 + 2x + 4).$$
$$\text{L.C.M.} = 3(x - 2)(x - 2)(x^2 + 2x + 4).$$

The factors have been written in "columns" in order to ease the determination of the number of times each factor must be used in the L.C.M.

9. Fractions. An algebraic fraction is the indicated quotient of two algebraic expressions. The simplest fraction is a/b in which a is called the **numerator** and b is called the **denominator.** The fraction

$1/x$ is known as the **reciprocal** of x. In *all* our discussion of fractions, the *denominator cannot be zero.* If for some values of the letters the denominator becomes zero, we say that the fraction is not defined there.

We shall start with the following properties of fractions:

1. *Multiplying, or dividing, both numerator and denominator of a fraction by the same number or expression does not change the value of the fraction.*

2. *Changing the signs of an* **even** *number of* **factors** *does* **not** *change the sign of the fraction, whereas changing the sign of an* **odd** *number of* **factors changes** *the sign of the fraction.*

Illustrations.

(1) $\dfrac{a+b}{a-b} = \dfrac{a+b}{a-b} \cdot \dfrac{a+b}{a+b} = \dfrac{(a+b)^2}{a^2-b^2}.$

(2) $\dfrac{1}{x-y} = \dfrac{-1}{-(x-y)} = -\dfrac{1}{-(x-y)} = -\dfrac{1}{y-x} = \dfrac{-1}{y-x}.$

(3) $\dfrac{(x-y)(a-b)}{(y-x)(b+a)} = -\dfrac{(x-y)(a-b)}{(x-y)(b+a)} = \dfrac{b-a}{b+a}.$

In considering the subject of algebraic fractions we are concerned with the four fundamental operations and the reduction of the fractions to their lowest terms. **Addition** and **subtraction** are accomplished by reducing the fractions to the lowest common denominator (L.C.D.) and then finding the algebraic sum of the resulting numerators:

$$\frac{a}{d} + \frac{b}{d} = \frac{a+b}{d} \quad \text{and} \quad \frac{a}{d} - \frac{b}{d} = \frac{a-b}{d}.$$

The finding of the L.C.D. of a set of fractions is easily accomplished by finding the L.C.M. of the denominators involved.

Illustrations.

(1) $\dfrac{3}{x-2} - \dfrac{4x+1}{x^2-4} + \dfrac{5}{x+2}.$

L.C.D. $= (x-2)(x+2)$; then:

$$\frac{3(x+2)}{(x-2)(x+2)} - \frac{4x+1}{(x-2)(x+2)} + \frac{5(x-2)}{(x-2)(x+2)}$$

$$= \frac{3x+6-4x-1+5x-10}{(x-2)(x+2)}$$

$$= \frac{4x-5}{(x-2)(x+2)}.$$

(2) $\dfrac{x-2}{x^2-x-6} + \dfrac{x+1}{x^2+x-2} + \dfrac{x+3}{x^2-4x+3}.$

L.C.D. $= (x-3)(x+2)(x-1)$; then:

$$\dfrac{(x-2)(x-1)+(x+1)(x-3)+(x+3)(x+2)}{(x-3)(x+2)(x-1)}$$

$$= \dfrac{x^2-3x+2+x^2-2x-3+x^2+5x+6}{(x-3)(x+2)(x-1)}$$

$$= \dfrac{3x^2+5}{(x-3)(x+2)(x-1)}.$$

The **product** of two fractions is the fraction whose numerator is the product of their numerators and whose denominator is the product of their denominators.

$$\frac{a}{b} \cdot \frac{x}{y} = \frac{ax}{by}.$$

To obtain the **quotient** of two fractions, invert the divisor and multiply.

$$\frac{a}{b} \div \frac{x}{y} = \frac{a}{b} \cdot \frac{y}{x} = \frac{ay}{bx}.$$

Illustrations.

(1) $\dfrac{a^2-b^2}{4a} \cdot \dfrac{8a}{a^2+2ab+b^2} = \dfrac{(a-b)(a+b)}{4a} \cdot \dfrac{\overset{2}{8a}}{(a+b)(a+b)}$

$$= \dfrac{2(a-b)}{a+b}.$$

(2) $\left[\dfrac{x-y}{16c} \div \dfrac{x^3+y^3}{8c^2}\right] \cdot \dfrac{x^2-xy+y^2}{c}$

$$= \dfrac{x-y}{16c} \cdot \dfrac{8c^2}{(x+y)(x^2-xy+y^2)} \cdot \dfrac{x^2-xy+y^2}{c}$$

$$= \dfrac{x-y}{2(x+y)}.$$

10. Simplification of Fractions. A fraction may become very complex if the numerator or denominator, or both, are fractions. In such cases it is desired to reduce them to their simplest form. In these complex fractions the place of the major division must be made clear; this is usually done by a heavy rule. The simplification is accomplished by applying the above principles in a systematic order. We shall illustrate with some examples.

Illustrations.

(1) Simplify $\dfrac{\dfrac{1-c}{c}+\dfrac{c}{1+c}}{\dfrac{c}{c+1}+\dfrac{c-1}{c}}$.

Solution.

$$\frac{\dfrac{1-c}{c}+\dfrac{c}{1+c}}{\dfrac{c}{c+1}+\dfrac{c-1}{c}}=\frac{\dfrac{(1-c)(1+c)+c^2}{c(1+c)}}{\dfrac{c^2+(c-1)(c+1)}{c(c+1)}}$$

$$=\frac{1-c^2+c^2}{c(1+c)}\cdot\frac{c(c+1)}{c^2+c^2-1}$$

$$=\frac{1}{2c^2-1}.$$

(2) Simplify $1-\dfrac{1}{2-\dfrac{1}{3-\frac14}}=N$.

Solution.

$$N=1-\frac{1}{2-\dfrac{1}{\dfrac{12-1}{4}}}$$

$$=1-\frac{1}{2-\dfrac{4}{11}}=1-\frac{1}{\dfrac{22-4}{11}}$$

$$=1-\frac{11}{18}=\frac{18-11}{18}=\frac{7}{18}.$$

11. Operations on Radicals. If radicals have the same index and the same radicand, they are called **like** radicals and may be added or subtracted as may other like expressions. If radicals are **unlike**, their algebraic sum can only be indicated. In order to perform the algebraic addition, we first reduce the radicals to their simplest form.

Illustrations.

(1) Find the sum of $2\sqrt{x}+\sqrt{4x}-\sqrt{36x^3}$.

Solution.

$$2\sqrt{x}+\sqrt{4x}-\sqrt{36x^3}=2\sqrt{x}+2\sqrt{x}-6x\sqrt{x}$$
$$=(4-6x)\sqrt{x}.$$

(2) Find the sum of $\sqrt{ax^2}-x\sqrt{4a}+3a\sqrt{25ax^2}$.

Solution.

$$\sqrt{ax^2} - x\sqrt{4a} + 3a\sqrt{25ax^2} = x\sqrt{a} - 2x\sqrt{a} + 15ax\sqrt{a}$$
$$= (15ax - x)\sqrt{a}.$$

The **product** of two radicals is obtained by applying the laws discussed in Section 6. Let us now illustrate with some examples, as follows.

Illustrations.

(1) Find the product $(\sqrt{15})(\sqrt{\frac{5}{3}})$.

Solution.

$$(\sqrt{15})(\sqrt{\tfrac{5}{3}}) = \sqrt{15(\tfrac{5}{3})} = \sqrt{25} = 5.$$

(2) Find the product $(\sqrt[3]{18})(\sqrt{6})$.

Solution.

$$(\sqrt[3]{18})(\sqrt{6}) = (18^{\frac{1}{3}})(6^{\frac{1}{2}}) = (18)^{\frac{2}{6}}(6)^{\frac{3}{6}}$$
$$= \sqrt[6]{18^2 \cdot 6^3} = \sqrt[6]{3^4 \cdot 2^2 \cdot 3^3 \cdot 2^3}$$
$$= \sqrt[6]{3^7 \cdot 2^5} = 3\sqrt[6]{3 \cdot 32}$$
$$= 3\sqrt[6]{96}.$$

(3) Find the product $(2\sqrt{x} - 3\sqrt{y})(7\sqrt{x} + 6\sqrt{y})$.

Solution.

$$
\begin{array}{r}
2\sqrt{x} - 3\sqrt{y} \\
7\sqrt{x} + 6\sqrt{y} \\
\hline
14x - 21\sqrt{xy} \\
+ 12\sqrt{xy} - 18y \\
\hline
14x - 9\sqrt{xy} - 18y
\end{array}
$$

The **division** of radicals usually results in the operation called **rationalizing the denominator,** that is, removing all radicals from the denominator. This is accomplished by multiplying both numerator and denominator by a rationalizing factor. For this purpose we recall the very useful product:

$$(a + b)(a - b) = a^2 - b^2.$$

Illustrations.

(1) Simplify $\sqrt{3} \div 7\sqrt{2}$.

Solution.

$$\frac{\sqrt{3}}{7\sqrt{2}} = \frac{\sqrt{3}}{7\sqrt{2}} \cdot \frac{\sqrt{2}}{\sqrt{2}} = \frac{\sqrt{6}}{7 \cdot 2} = \frac{1}{14}\sqrt{6}.$$

(2) Simplify $(\sqrt{5} + x) \div (\sqrt{5} - x)$.

Solution.

$$\frac{\sqrt{5} + x}{\sqrt{5} - x} = \frac{\sqrt{5} + x}{\sqrt{5} - x} \cdot \frac{\sqrt{5} + x}{\sqrt{5} + x} = \frac{(\sqrt{5} + x)^2}{5 - x^2}$$

$$= \frac{5 + x^2 + 2x\sqrt{5}}{5 - x^2}.$$

(3) Simplify $(\sqrt[6]{c} \div \sqrt[4]{c^2}) \cdot \sqrt{c^2}$.

Solution.

$$(\sqrt[6]{c} \div \sqrt[4]{c^2})(\sqrt{c^2}) = c^{\frac{1}{6}} \cdot c^{-\frac{2}{4}} \cdot c^{\frac{2}{2}}$$

$$= c^{\frac{1}{6} - \frac{1}{2} + 1} = c^{\frac{4}{6}} = c^{\frac{2}{3}}$$

$$= \sqrt[3]{c^2}.$$

12. Imaginary Numbers. We have seen that the cube root of -8 has a meaning and a value, -2. But what about the square root of -4? Is there a number such that when multiplied by itself the result is -4? In the real-number system there is not; neither $+2$ nor -2 will yield the desired result. The operation of extracting an *even root of a negative number* requires us to define a new number which is called an **imaginary number.** It is a number which has the property that when multiplied by itself it yields a negative number. We let i denote the **imaginary unit** $\sqrt{-1}$; then we have:

$$i^2 = (\sqrt{-1})^2 = -1.$$

This permits us to write the square root of a negative number as the product of a real number and the imaginary number i, thus:

$$\pm\sqrt{-4} = \pm i\sqrt{4} = \pm 2i.$$

Suppose we raise the number i to successive positive integral powers; we obtain:

$$
\begin{aligned}
i &= \sqrt{-1}; & i^5 &= i; \\
i^2 &= -1; & i^6 &= -1; \\
i^3 &= -i; & i^7 &= -i; \\
i^4 &= 1; & i^8 &= 1.
\end{aligned}
$$

Thus we see that the integral powers of i can be reduced to i, -1, $-i$, or 1.

The combination of an imaginary number and a real number of the form $a + bi$ is called a complex number. We shall discuss complex numbers in Chapter IV. At the present we are interested in simplifying expressions containing the imaginary quantity.

Illustrations.

(1) Simplify $i^5 + 3i^4 - 5i^3 + 6i^2 - 3i$.

Solution.

$$i^5 + 3i^4 - 5i^3 + 6i^2 - 3i = i + 3 + 5i - 6 - 3i$$
$$= 3i - 3.$$

(2) Simplify $\dfrac{2 - \sqrt{-3}}{2 + \sqrt{-3}}$.

Solution.

$$\frac{2 - \sqrt{-3}}{2 + \sqrt{-3}} = \frac{2 - i\sqrt{3}}{2 + i\sqrt{3}} \cdot \frac{2 - i\sqrt{3}}{2 - i\sqrt{3}}$$

$$= \frac{4 - 4i\sqrt{3} + 3i^2}{4 - 3i^2}$$

$$= \frac{4 - 4i\sqrt{3} - 3}{4 + 3} = \frac{1 - 4i\sqrt{3}}{7}.$$

13. Mistakes. We shall close this chapter with a list of common mistakes made by students when working with the subject matter of this chapter. We shall indicate that two quantities are not equal ($\neq$). Students, however, make the error of indicating that the quantities are equal. It is suggested that the student read this list, understand why the $\neq$ is used, and also correct the errors.

1. $|-3| \neq -3$.
2. $3^2 \cdot 3^3 \neq 9^5$.
3. $a^2 \cdot b^5 \neq (ab)^7$.
4. $x + y - 3(z + w) \neq x + y - 3z + w$.
5. $a + (x - 3y) \neq a + x + 3y$.
6. $3a + 4b \neq 7ab$.
7. $7 - (-3) \neq 7 - 3$.
8. $3x^{-1} \neq \dfrac{1}{3x}$.
9. $\sqrt{x^2 + y^2} \neq x + y$.
10. $\dfrac{x + y}{x + z} \neq \dfrac{y}{z}$.
11. $\dfrac{1}{x - y} \neq -\dfrac{1}{x + y}$.
12. $\dfrac{x}{y} + \dfrac{r}{s} \neq \dfrac{x + r}{y + s}$.
13. $x\left(\dfrac{a}{b}\right) \neq \dfrac{ax}{bx}$.
14. $\dfrac{xa + xb}{x + xd} \div x \neq \dfrac{a + b}{1 + d}$.
15. $\sqrt{-x}\sqrt{-y} \neq \sqrt{xy}$.

14. Exercise I.

1. Simplify:
   (a) $8x - (14y + 3x) - \{2x - (4y + x)\} - (4y - x)$.
   (b) $-\frac{1}{2} - \{\frac{3}{4} - 3[\frac{1}{3} - (\frac{3}{4} - \frac{7}{6})]\}$.

2. Perform the indicated operations:
   (a) $(-2ab^2c^3)^3$.

   (b) $\dfrac{42a^2(x^3y)^4}{14(ax)^2y^3}$.

   (c) $\left(\dfrac{x^{2n-1}}{y^{m+1}}\right)^3 \div \left(\dfrac{x^{n+1}}{y^m}\right)$.

   (d) $\sqrt[4]{a^2 - 2ax + x^2}$.

   (e) $\sqrt{2} - \sqrt{8} + \sqrt{32} - \sqrt{128}$.

3. Multiply $4x^2 - 3xy + 2y^2$ by $x^2 - 4xy - 3y^2$.

4. Divide $4x^4 - 19x^3y + 2x^2y^2 + xy^3 - 6y^4$ by $-4xy - 3y^2 + x^2$.

5. Factor the following expressions:
   (a) $81ax^3 - 27bx^3 + 8by^3 - 24ay^3$.
   (b) $a^4 - 81b^4$.
   (c) $6x^2 + 19xy + 10y^2$.

6. Combine into one fraction and simplify:

$$\frac{x + y}{6x^2 + 19xy + 10y^2} - \frac{2x - 5y}{3x^2 - xy - 2y^2} - \frac{3x - 2y}{2x^2 + 3xy - 5y^2}.$$

7. Simplify:

$$\frac{x}{1 - \dfrac{1}{1 - x}} - \frac{x^2}{x - \dfrac{x}{1 - x}}.$$

8. Multiply $\sqrt{x} + 3\sqrt{y}$ by $\sqrt{x} - 3\sqrt{y}$.

9. Rationalize the denominator:

$$\frac{\sqrt{a + b} - 5\sqrt{a - b}}{3\sqrt{a + b} - 2\sqrt{a - b}}.$$

10. Simplify:
    (a) $\sqrt{-2}\,\sqrt{-3}\,\sqrt{-4}\,\sqrt{-5}$.

    (b) $\dfrac{3 - \sqrt{-2}}{3 + \sqrt{-2}}$.

2

FUNCTIONS AND THEIR REPRESENTATION

15. Constants and Variables. We have already discussed the use of letters to represent quantities in a discussion, and we are already familiar with the concept of a formula. In the formulas some of the symbols represent fixed numbers, whereas others may be assigned values at will. This leads to the following definitions.

A symbol which, throughout a discussion, represents a fixed number is called a **constant.**

A symbol which, throughout a discussion, may assume different values is called a **variable.**

Illustration. Consider the formula for the area of a circle, $A = \pi r^2$. The number π is constant; the quantities A and r are variables.

16. Functions and Notation. One of the most basic concepts in mathematics is the concept of function. Let us first illustrate the concept by considering the example of the area of a circle, $A = \pi r^2$, where A and r are variables. We have written here a relationship between A and r such that if values are given to r we can determine corresponding values of A. This gives us an insight to a most useful definition of function.*

Definition. *If two variables* x *and* y *are so related that, for a value assigned to* x, *one or more values of* y *are determined, then* y *is said to be a* **function of** x.

In order to designate that y is a function of x we employ a **functional notation:**

$$y = f(x)$$

which is read "y equals f of x." The expression $f(x)$ does *not* mean f times x when used in this sense. In this relationship x is called the

---

* The student is urged to read the treatment of function given in Section 150, Chapter 15.

independent variable and y, the **dependent variable**, since y is determined after a value has been assigned to x. There are many examples of functions:

1. A linear function, $y = f(x) = ax + b$;
2. A quadratic function, $y = f(x) = ax^2 + bx + c$;
3. A formula for distance, $d = f(v,t) = vt$;

and many others.

It is possible to have a function of more than one variable such as the formula for distance, $d = vt$, where d depends upon both v and t for its values. We then say that d is a function of v and t, and it is written $f(v,t)$.

To obtain the value of a function for given values of the independent variable, substitute the given value into the function and perform the expressed arithmetic.

Illustration. Find the values of the function:

$$f(x) = x^3 - 2x^2 + x$$

for $x = 1, 2, -1, 0$, and a.

Solution.

$$f(1) = (1)^3 - 2(1)^2 + (1) = 1 - 2 + 1 = 0.$$
$$f(2) = (2)^3 - 2(2)^2 + (2) = 8 - 8 + 2 = 2.$$
$$f(-1) = (-1)^3 - 2(-1)^2 + (-1) = -1 - 2 - 1 = -4.$$
$$f(0) = (0)^3 - 2(0)^2 + (0) = 0.$$
$$f(a) = a^3 - 2a^2 + a.$$

We shall now introduce another mathematical notation, Δx, read "delta x." In this case we do *not* mean Δ times x but consider Δx as *one quantity*, just as $f(x)$ is used as one symbol. The symbol Δx is used when we are varying an independent variable, and denotes the **increment** by which the variable is changed. If we have $y = f(x)$ and change x, then there is a corresponding change in y and we call Δy the increment of y. The value of the increment can be obtained by subtracting two consecutive values of the variable.

Illustration 1. Let $y = 2x + 1$ and assign values to x as shown in the table on the opposite page. We compute values of y and the increments.

If we let $y = f(x)$ and add an increment to x, we have:

$$y + \Delta y = f(x + \Delta x)$$

and we can find Δy by subtracting y from both sides of the equation; thus we have:

$$\Delta y = f(x + \Delta x) - y = f(x + \Delta x) - f(x).$$

x	Δx	y	Δy
1		3	
	1		2
2		5	
	1		2
3		7	
	2		4
5		11	
	1		2
6		13	
	3		6
9		19	

This is a very important expression in advanced mathematics.

We wish to emphasize that when we write $f(x + \Delta x)$ we are replacing x by $x + \Delta x$ everywhere in the function.

Illustration 2. Let $f(x) = 3x^2 + x$.

Then:

$$f(x + \Delta x) = 3(x + \Delta x)^2 + (x + \Delta x)$$
$$= 3[x^2 + 2x\Delta x + (\Delta x)^2] + x + \Delta x$$
$$= 3x^2 + (6\Delta x + 1)x + 3(\Delta x)^2 + \Delta x.$$

The use of incremental changes leads to a most useful definition.

Definition. *The* **average rate of change** *of a variable y with respect to a variable x within a given interval is the quotient* $\Delta y/\Delta x$.

If we consider time as the independent variable, this is the definition of the **average velocity** of the dependent variable.

We note that the average rate of change of y in Illustration 1 at each point is:

$\dfrac{\Delta y}{\Delta x}$	$\dfrac{2}{1} = 2$	$\dfrac{2}{1} = 2$	$\dfrac{4}{2} = 2$	$\dfrac{2}{1} = 2$	$\dfrac{6}{3} = 2$

The rate of change of y is a constant, 2, in this example. The fact that rate of change is a constant is characteristic of a linear function.

17. Coordinate System. In order to give a pictorial representation to functions, mathematicians have devised a coordinate system which describes the position of a point in a plane. To define this coordinate system let us consider two number scales as pictured in Fig. 1, and let them intersect each other at right angles at the point O. We have now formed a **rectangular coordinate system,** and the two lines are called the **axes,** denoted by OX and OY. See Fig. 2. Let P be any point in the plane OX and OY.

The **horizontal coordinate** of P is the directed perpendicular distance, x, from OY to P. It is said to be **positive** when P is to the *right* of OY and **negative** when P is to the *left* of OY. It is called the **abscissa.**

The **vertical coordinate** of P is the directed perpendicular distance, y, from OX to P. It is said to be **positive** when P is *above* OX and **negative** when P is *below* OX. It is called the **ordinate.**

The abscissa and ordinate together are called the **coordinates** of P and are denoted by (x,y). The point O is called the **origin,** and OP is called the **radius vector, r;** r is zero if P coincides with O; otherwise it is a positive quantity.

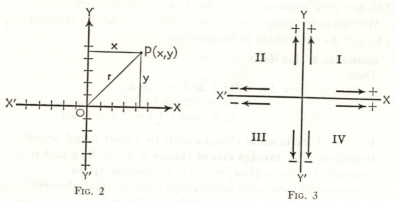

Fig. 2   Fig. 3

The coordinate axes divide the plane into four **quadrants** and these are numbered counterclockwise I, II, III, and IV. The point P may be located on this plane by simply knowing the values of its coordinates x and y. We see that in the first quadrant both x and y are positive; in the second quadrant x is negative, y positive; in the third quadrant $(-,-)$; and in the fourth quadrant $(+,-)$; see Fig. 3.

To **plot** a point, $P(x,y)$, is to locate its position from the values of x and y.

Illustration. Plot the points $P_1(2,1)$, $P_2(-3,2)$, $P_3(-2,-4)$, $P_4(3,-3)$. *Solution.* See Fig. 4.

The coordinate system we have defined is usually called a *Cartesian coordinate system.*

18. Graph of a Function; Locus. We can now give a pictorial representation of a function. This is called a **graph** or **locus** of a function. It is obtained by plotting arbitrarily assigned values of the independent variable as abscissas and the corresponding computed values of the function as ordinates. The first step is to obtain a table of values of the function for chosen values of the independent variable. Next, these pairs of values are plotted as points. The points are then joined by a smooth curve, and we have a picture (graph) of the function. In order to ease the plotting, graph paper is usually used. It is paper which has a square grid drawn on it and thus forms a natural coordinate system when we simply choose the origin, axes, and unit length.

Illustration. Plot the graph of the function:

$$y = 3x - 4.$$

Solution. Calculate the table of values.

x	-1	0	1	2	3
y	-7	-4	-1	2	5

Plot the corresponding points and join them with a curve; see Fig. 5.

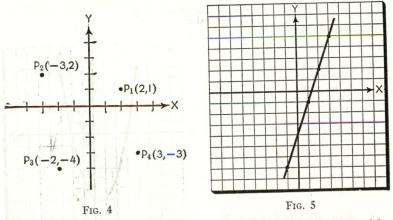

FIG. 4   FIG. 5

It is to be noted that as many points as desired can be plotted by simply assigning more values to x and calculating the corresponding values of y from the function.

The term *locus* is an important term in mathematics, and although we have linked it with the term *graph*, it merits a more rigorous definition.

Definition. *The* **locus** *of an equation is the totality of all points, and only those points, whose coordinates satisfy the equation.*

To understand this statement let us clearly focus the two things involved, namely:

1. Every point whose coordinates satisfy the equation lies on the locus.

2. The coordinates of every point on the locus satisfy the equation.

The locus is also thought of as the *path of a point*, $P(x,y)$, *which moves according to a specified law*.

Let us now return to the term *graph* and give a more rigorous definition.

Definition. *The* **graph** *of the equation* $y = f(x)$ *is the totality (aggregate) of all points whose coordinates* (x,y) *satisfy the relation* $y = f(x)$.

19. Zero of a Function; Intercepts. If a function is equal to zero for special values of the independent variable, the values are called **zeros** of the function. To explain further, let $y = f(x)$; then the zeros of this function are those values of x for which $y = 0$. If we draw the graph of $y = f(x)$, the zeros are the x values or the abscissas of the point where the graph crosses (or touches) the X-axis.

Illustration. Plot the graph of the function:

$$y = f(x) = x^2 + x - 6$$

and find the zeros.
Solution.

x	y
0	-6
1	-4
2	0
3	6
-1	-6
-2	-4
-3	0
-4	6

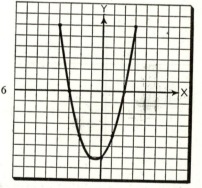

Fig. 6

The zeros are $x = 2$, $x = -3$.

In geometry the value of x for which the curve crosses the X-axis is called the **x-intercept**. Similarly, the value of y for which the curve crosses the Y-axis is called the **y-intercept**. To obtain these intercepts we set y equal to zero and solve for x; then set x equal to zero and solve for y.

In the above illustration the x-intercepts are the zeros of the function, $x = 2$, $x = -3$. The y-intercept is $y = -6$.

20. Distance between Two Points. The distance between two points in a plane may be found from their coordinates. Let the two points be $P_1(x_1, y_1)$ and $P_2(x_2, y_2)$. Plot these points on a coordinate system; see Fig. 7. Form the right triangle P_1RP_2 and apply the Theorem of Pythagoras * to obtain:

$$d^2 = \overline{P_1R}^2 + \overline{RP_2}^2.$$

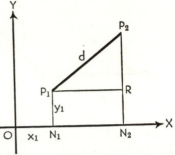

We note that:

$$P_1R = N_1N_2 = ON_2 - ON_1 = x_2 - x_1;$$
$$RP_2 = N_2P_2 - N_2R = y_2 - y_1,$$

and substituting these values into the above formula, we have, after taking the square root of both sides:

$$d = \sqrt{(x_2 - x_1)^2 + (y_2 - y_1)^2}.$$

FIG. 7

This is the formula for the distance between two points in terms of the coordinates of the points.**

Illustration. Find the distance between the following points:

(1) $P_1(2,3)$ and $P_2(4,1)$.

Solution.

$$d = \sqrt{(4 - 2)^2 + (1 - 3)^2}$$
$$= \sqrt{4 + 4} = 2\sqrt{2}.$$

(2) $P_1(-4,3)$ and $P_2(2,-1)$.

Solution.

$$d = \sqrt{(2 + 4)^2 + (-1 - 3)^2}$$
$$= \sqrt{36 + 16} = 2\sqrt{13}.$$

(3) $P_1(2,5)$ and $P_2(-4,-3)$.

Solution.

$$d = \sqrt{(-4 - 2)^2 + (-3 - 5)^2}$$
$$= \sqrt{36 + 64} = 10.$$

* "The square on the hypotenuse of a right triangle is equal to the sum of the squares on the other two sides."

** In this formula we have tacitly taken the "+" sign with the radical in order to preserve the physical concept of a distance being a positive quantity. The negative radical also exists, and had we employed it, we would be making a distinction between the distance from P_1 to P_2 and the distance from P_2 to P_1. Thus we would have directed distances, which is sometimes convenient.

21. The General Angle. The reader is already familiar with the definition of an angle. We shall discuss the angle in its most general sense. Let any line segment coinciding with OA (Fig. 8) be revolved about one of its end points, say O, until it takes the position OB.

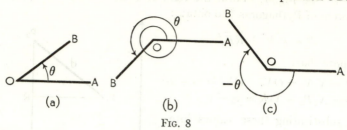

Fig. 8

The revolution of the line has generated the angle AOB. The line may be revolved about O in a clockwise or counterclockwise direction, and there may be no limit to the number of times it is revolved. This has defined a general angle; we shall speak of it as being *positive* if the line is moved in a counterclockwise direction and as *negative* if moved in a clockwise direction. The line OA is said to be the **initial side** and OB, the **terminal side.**

The general angle may be measured in the usual sexagesimal measure or in the circular (radian) measure.*

A. Sexagesimal Measure.

The *degree* $= \frac{1}{360}$ of one revolution $= 1°$;
the *minute* $= \frac{1}{60}$ of one degree $= 1'$;
the *second* $= \frac{1}{60}$ of one minute $= 1''$.

B. Radian Measure.

A *radian* is defined to be the central angle subtended by an arc of a circle equal in length to the radius of the circle.

C. Conversion Units.

The above systems of measurement are related by:

$$1 \text{ degree} = \frac{\pi}{180} \text{ radians};$$

$$1 \text{ radian} = \frac{180}{\pi} \text{ degrees.}$$

---

* There is still a third measure known as the **mil** which is used by the military. For a discussion see K. L. Nielsen and J. F. Heyda, *The Mathematical Theory of Airborne Fire Control*, Government Printing Office, Washington, 1951, p. 182.

Thus, *to change from degrees to radians* multiply the number of degrees by $\pi/180$; *to change from radians to degrees* multiply the number of radians by $180/\pi$.

In decimal form we have:

$$1° = 0.017453 \text{ radians};$$
$$1 \text{ rdn.} = 57.2958°.$$

Illustrations.

(1) Change $\pi/6$ radians to degrees.

Solution.

$$\frac{\pi}{6}\left(\frac{180}{\pi}\right) = 30°.$$

(2) Change 45° to radian measure.

Solution.

$$45\left(\frac{\pi}{180}\right) = \frac{\pi}{4} \text{ radians.}$$

The definition of a radian leads directly to a formula for the length of arc on a circle. Since the definition says:

$$\text{angle in radians} = \frac{\text{arc}}{\text{radius}}; \quad \theta = \frac{s}{r} \ (\theta \text{ in radians}),$$

we have:

$$s = r\theta.$$

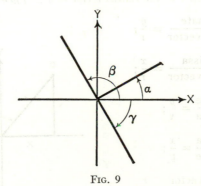

Fig. 9

If we take a general angle and place it so that O is at the origin of the rectangular coordinate plane and its initial side coincides with the positive X-axis, the angle is said to be in its **standard position**. Fig. 9 shows the three angles:

$$\alpha = 30°; \quad \beta = 120°; \quad \gamma = -60°.$$

22. Trigonometric Functions. We shall now define a special set of functions. They follow the general definition of a function in that they express a relationship between two variables. We shall first specialize by considering the independent variable to be an angle. There are six main functions which are known as the **trigonometric functions:**

NAME OF FUNCTION	ABBREVIATION
sine of θ	$\sin \theta$
cosine of θ	$\cos \theta$
tangent of θ	$\tan \theta$
cotangent of θ	$\cot \theta$
secant of θ	$\sec \theta$
cosecant of θ	$\csc \theta$

The reader may already be familiar with these functions and may have had them defined for an acute angle (less than 90°) in terms of the sides of a right triangle. We shall define them for the general angle.

Definitions. *Let θ be any angle in its standard position. Let P be any point (not the origin) having coordinates (x,y) and lying on the terminal side of θ. Let r be the radius vector of P. Then:*

$$\sin \theta = \frac{\text{ordinate}}{\text{radius vector}} = \frac{y}{r};$$

$$\cos \theta = \frac{\text{abscissa}}{\text{radius vector}} = \frac{x}{r};$$

$$\tan \theta = \frac{\text{ordinate}}{\text{abscissa}} = \frac{y}{x};$$

$$\cot \theta = \frac{\text{abscissa}}{\text{ordinate}} = \frac{x}{y};$$

$$\sec \theta = \frac{\text{radius vector}}{\text{abscissa}} = \frac{r}{x};$$

$$\csc \theta = \frac{\text{radius vector}}{\text{ordinate}} = \frac{r}{y}.$$

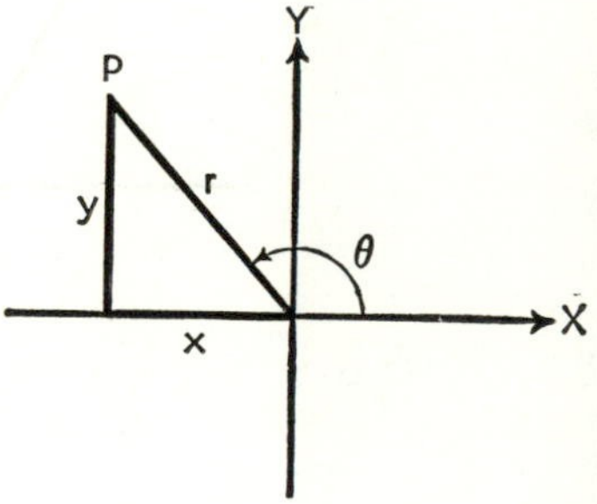

FIG. 10

The student should commit these definitions to memory.

If the angle θ is acute, the terminal side falls in the first quadrant and we have a right triangle. The definitions can then be specialized to the sides of a right triangle.*

It is immaterial to the definition of the trigonometric functions where the point P falls on the radius vector, r, since by similar triangles we can prove that the ratios for a given angle are constant and the trigonometric functions of given angles are fixed numbers. Of special interest are the trigonometric functions of certain angles, and we shall place them in the following table.

	SIN	COS	TAN	COT	SEC	CSC
0°	0	1	0	—	1	—
30°	$\frac{1}{2}$	$\frac{1}{2}\sqrt{3}$	$\frac{1}{3}\sqrt{3}$	$\sqrt{3}$	$\frac{2}{3}\sqrt{3}$	2
45°	$\frac{1}{2}\sqrt{2}$	$\frac{1}{2}\sqrt{2}$	1	1	$\sqrt{2}$	$\sqrt{2}$
60°	$\frac{1}{2}\sqrt{3}$	$\frac{1}{2}$	$\sqrt{3}$	$\frac{1}{3}\sqrt{3}$	2	$\frac{2}{3}\sqrt{3}$
90°	1	0	—	0	—	1
180°	0	−1	0	—	−1	—
270°	−1	0	—	0	—	−1

The values are obtained directly from the definition by giving corresponding values to x, y, and r. For example, place the angle 45° in standard position and let $x = y = 1$; then $r = \sqrt{2}$; see Fig. 11. From the definitions we have:

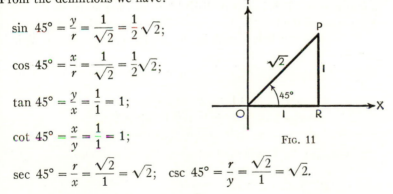

$$\sin 45° = \frac{y}{r} = \frac{1}{\sqrt{2}} = \frac{1}{2}\sqrt{2};$$

$$\cos 45° = \frac{x}{r} = \frac{1}{\sqrt{2}} = \frac{1}{2}\sqrt{2};$$

$$\tan 45° = \frac{y}{x} = \frac{1}{1} = 1;$$

$$\cot 45° = \frac{x}{y} = \frac{1}{1} = 1;$$

Fig. 11

$$\sec 45° = \frac{r}{x} = \frac{\sqrt{2}}{1} = \sqrt{2}; \quad \csc 45° = \frac{r}{y} = \frac{\sqrt{2}}{1} = \sqrt{2}.$$

* In terms of the sides of a right triangle:

$$\sin \theta = \frac{\text{opposite side}}{\text{hypotenuse}}; \quad \cos \theta = \frac{\text{adjacent side}}{\text{hypotenuse}}; \quad \tan \theta = \frac{\text{opposite side}}{\text{adjacent side}};$$

$$\cot \theta = \frac{\text{adjacent side}}{\text{opposite side}} \quad \sec \theta = \frac{\text{hypotenuse}}{\text{adjacent side}} \quad \csc \theta = \frac{\text{hypotenuse}}{\textbf{opposite side}}.$$

As another example, consider the angle 180°. A point P on the terminal side will have $(-x_1, 0)$ for its coordinates and $r = \sqrt{0 + x_1{}^2} = x_1$ so that we have:

$$x = -x_1, \quad y = 0, \quad r = x_1,$$

and by the definitions:

$$\sin 180° = \frac{y}{r} = \frac{0}{x_1} = 0; \qquad \cot 180° = \frac{x}{y} = \frac{-x_1}{0}, \text{ undefined};$$

$$\cos 180° = \frac{x}{r} = \frac{-x_1}{x_1} = -1; \qquad \sec 180° = \frac{r}{x} = \frac{x_1}{-x_1} = -1;$$

$$\tan 180° = \frac{y}{x} = \frac{0}{-x_1} = 0; \qquad \csc 180° = \frac{r}{y} = \frac{x_1}{0}, \text{ undefined}.$$

The reader should find all the values of the table in a similar manner.

23. Trigonometric Tables.* Since the trigonometric functions of a given angle are constants, they may be computed and placed in a table. The values are usually expressed in decimal form correct to a given number of places. There are many such tables; however, we shall consider an abbreviated four-place table in order to learn the theory and use of such tables.

Before turning to the tables proper let us first consider the matter of expressing a number correct to a certain number of places. Since, in general, these numbers are endless decimals, we must "round off" to four places. This is accomplished by throwing away all the digits to the right of the fourth place, and if this discarded number is:

(a) greater than half a unit in the 4th place, increase the digit in that place by 1;

(b) less than half a unit in the 4th place, leave the digit in that place unaltered;

(c) exactly half a unit in the 4th place,

(i) increase an *odd* digit in the 4th place by 1;

(ii) leave an *even* digit in the 4th place unaltered.

Illustrations. Express the following numbers to four places:

(1) 64.347. Answer: 64.35.
(2) 13.342. Answer: 13.34.
(3) 7.2345. Answer: 7.234.
(4) 7.2375. Answer: 7.238.

---

* For a complete discussion of more extensive tables see K. L. Nielsen and J. H. VanLonkhuyzen, *Plane and Spherical Trigonometry*, Barnes & Noble, New York, 1957, pp. 8–12.

Table V in the back of this book is a four-place table of the trigonometric functions of the acute angles at every 10 minutes. The angles from 0° to 45° are labeled in the left-hand column, and the angles from 45° to 90° in the extreme right-hand column. The trigonometric functions are labeled at the top and the bottom. For the angles in the *left*-hand column we read the titles of the columns at the *top* of the page. For those in the extreme *right*-hand column we read the titles of the columns at the *bottom* of the page.

Let us now consider the following operations.

I. *Given the angle; to find the trigonometric function.*

Illustrations.

(1) To find tan 13° 40′.

Solution. Find 13° in the extreme left column. Find 40′ under 13°, and opposite this in the column headed by "tan" at the *top* of the page read .2432. Thus tan 13° 40′ = .2432.

(2) To find cos 79° 20′.

Solution. Find 79° in the extreme right column. Find 20′ *above* 79° in this column, and opposite 20′ in the column headed by "cos" at the *bottom* of the page read .1851. Thus cos 79° 20′ = .1851.

II. *Given the trigonometric function; to find the angle.*

Illustrations.

(1) Given sin θ = .3283; find θ.

Solution. In the columns headed by "sin" search for .3283. This is found in the column which is headed by "sin" at the *top* of the page; therefore, look in the *left*-hand column and read 19° 10′. Thus θ = 19° 10′.

(2) Given cot θ = .4950. Find θ.

Solution. In the columns headed by "cot" search for .4950. This is found in the column headed by "cot" at the *bottom* of the page; therefore, look in the *right*-hand column for the angle and read 63° 40′.

Note: In the *left* columns of the table the angle increases as one reads *downward;* in the *right* columns, as one reads *upward.*

III. *Interpolation.* If it is desired to obtain the trigonometric functions of an angle given to the nearest minute, it becomes necessary to interpolate for the values. This is done either by linear interpolation or by using tables of proportional parts. Let us discuss linear interpolation, which will suffice for our work. The difference between two entries in the table is called a **tabular difference.** Interpolation is accomplished by taking appropriate portions of this difference.

Illustrations.

(1) Find tan 49° 33′.

Solution. Since tan 49° 33′ is between tan 49° 30′ and tan 49° 40′, we obtain these values and subtract one from the other.

$$\begin{array}{r} \tan 49° 40' = 1.1778 \\ \underline{\tan 49° 30' = 1.1708} \\ \text{tabular difference} = .0070 \end{array}$$

The tabular difference of the function is now multiplied by the difference between the given angle and the smaller chosen angle and then divided by the tabular difference of the angle. Thus, $(.0070)(33 - 30)/10 = (.0070)(.3) = .0021$. This amount is now added to the tangent of the smaller angle (49° 30′) and rounded off to four decimal places; the result is 1.1729.

Thus: tan 49° 33′ = 1.1729.

The work may be arranged as follows:

$$\left.\begin{array}{l} \tan 49° 30' = 1.1708 \\ \tan 49° 33' = 1.1729 \\ \tan 49° 40' = 1.1778 \end{array}\right] .0070(.3) = .0021.$$

The difference between the angles in this table is a constant 10′, and the division by 10 is a simple matter of inserting the decimal point.

(2) Find cos 32° 47′.

Solution.

$$\left.\begin{array}{l} \cos 32° 40' = .8418 \\ \cos 32° 47' = .8408 \\ \cos 32° 50' = .8403 \end{array}\right] (-.0015)(.7) = -.00105.$$

(3) Given tan θ = .8312. Find θ.

Solution. Search the table in the column headed "tan" for the value closest to .8312. We find:

$$\left.\begin{array}{l} \tan 39° 40' = .8292 \\ \tan \theta = .8312 \\ \tan 39° 50' = .8342 \end{array}\right] \tfrac{20}{50}(10) = 4.$$

Subtract the value of the trigonometric function of the smaller angle from that of the larger angle and also from the given value. Form the fraction of these differences and multiply by 10 (the angular difference). The result rounded off to the nearest integer is the number of minutes to be added to the *smaller* angle. Thus:

$$\theta = 39° 40' + 4' = 39° 44'.$$

It is not necessary to record the values of secant and cosecant since computations may be accomplished without these functions by the use of the other four.

24. Reduction to Acute Angles. In the tables discussed in the last section, we recorded only the values for angles from 0° to 90°. To obtain the functions of other angles we must be able to express them by means of the functions of the acute angle. First, let us determine the signs of the trigonometric functions in the various quadrants. The values of x and y may be either positive or negative; the value of r is always positive. Then in the first quadrant x and y are both positive and all the functions are positive. In the second quadrant x is negative and y is positive so that all the functions are negative except $\sin \theta$ and $\csc \theta$. We arrive at this by simply testing in the definitions of the functions.

$$\sin \theta = \frac{y}{r} = \frac{+}{+} = +; \quad \cos \theta = \frac{x}{r} = \frac{-}{+} = -.$$

Continuing this process through all the functions and in the other quadrants leads us to the following summary chart which should be committed to memory:

$$
\begin{array}{c|c}
\left.\begin{array}{l}\sin \theta \\ \csc \theta \end{array}\right\} + & \\
\left.\begin{array}{l}\text{all} \\ \text{others}\end{array}\right\} - & \text{all} \quad \Big\} + \\
\hline
\left.\begin{array}{l}\tan \theta \\ \cot \theta \end{array}\right\} + & \left.\begin{array}{l}\cos \theta \\ \sec \theta \end{array}\right\} + \\
\left.\begin{array}{l}\text{all} \\ \text{others}\end{array}\right\} - & \left.\begin{array}{l}\text{all} \\ \text{others}\end{array}\right\} -
\end{array}
$$

We shall now define a **reference angle** *as the acute angle α between the terminal side of θ and the X-axis.*

Theorem. *Any function of an angle θ, in any quadrant, is numerically equal to the same function of the reference angle for θ; i.e.,*

(any function of θ) $= \pm$ (same function of α).

The "+" or "−" is determined by the quadrant in which the angle falls.

Proof. Let θ be any angle in its standard position. On the terminal side of θ pick a point P, and drop a perpendicular to the X-axis.

(See Fig. 12.) The reference angle α is an acute angle of the right triangle ORP and we shall apply the right triangle definition to the functions of α; thus:

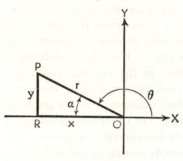

FIG. 12

$$\sin \alpha = \frac{RP}{OP}, \quad \cos \alpha = \frac{OR}{OP}, \text{ etc.}$$

Now $x = \pm OR$, $y = \pm RP$, and $r = OP$, the "$+$" or "$-$" depending upon the quadrant in which P falls. From the definition we have:

$$\sin \theta = \frac{y}{r} = \pm \frac{RP}{OP} = \pm \sin \alpha.$$

Similarly, for the other functions.

To obtain the reference angle α simply:*

$$\alpha = 180° - \theta \quad \text{for} \quad 90° < \theta < 180°;$$
$$\alpha = \theta - 180° \quad \text{for} \quad 180° < \theta < 270°;$$
$$\alpha = 360° - \theta \quad \text{for} \quad 270° < \theta < 360°.$$

For $\theta > 360°$, we have:

any function of θ = same function of $(\theta - n360°)$

where n is an integer.

Illustrations.

(1) Find $\sin 234° 10'$.

Solution. $\sin 234° 10' = -\sin (234° 10' - 180°)$
$= -\sin 54° 10' = -.8107.$

(2) Find $\cos 318° 40'$.

Solution. $\cos 318° 40' = \cos (360° - 318° 40')$
$= \cos 41° 20' = .7509.$

(3) Find $\tan 519° 20'$.

Solution. $\tan 519° 20' = \tan (519° 20' - 360°)$
$= \tan 159° 20'$
$= -\tan (180° - 159° 20')$
$= -\tan 20° 40' = -.3772.$

The property that:

any function of $(\theta + n360°)$ = same function of θ

* The notation $90° < \theta < 180°$ means "values of θ between 90° and 180°."

places the trigonometric functions in a special class. Any such function, which repeats its values at a regular interval, is said to be **periodic,** and the length of the interval is called its **period.** All the trigonometric functions are periodic; the tangent and cotangent repeat themselves every 180°; thus they have a period of 180° or π radians; the other four trigonometric functions have a period of 360° or 2π radians. It should now be clear that all the information about the values of the trigonometric functions can be obtained by considering the angles from 0° to 360°.

25. Functions of a Negative Angle. To obtain the functions of a negative angle we have a theorem.

Theorem. *If θ is any nonquadrantal angle, then:*

$$\sin (-\theta) = -\sin \theta; \quad \cot (-\theta) = -\cot \theta;$$
$$\cos (-\theta) = \cos \theta; \quad \sec (-\theta) = \sec \theta;$$
$$\tan (-\theta) = -\tan \theta; \quad \csc (-\theta) = -\csc \theta.$$

Proof. Place the angle $-\theta$ in its standard position. Construct the angle θ, numerically equal to $-\theta$, in its standard position. On the terminal sides of θ and $-\theta$ choose P_1 and P_2 so that $OP_1 = OP_2$. Then since $|\theta| = |-\theta|$, $P_1P_2 \perp OX$ and is bisected by OX. We then have (see Fig. 13):

$$r_1 = r_2, \quad x_1 = x_2, \quad y_1 = -y_2.$$

From the definition, we have:

$$\sin (-\theta) = \frac{y_2}{r_2} = \frac{-y_1}{r_1} = -\sin \theta;$$

$$\cos (-\theta) = \frac{x_2}{r_2} = \frac{x_1}{r_1} = \cos \theta;$$

$$\tan (-\theta) = \frac{y_2}{x_2} = \frac{-y_1}{x_1} = -\tan \theta;$$

etc.

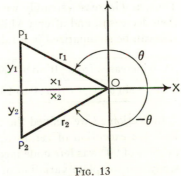

Fig. 13

To find the functions of a negative angle: **first** *change to the function of the corresponding positive angle by using the above theorem and then find the function of the positive angle.*

Illustrations.

(1) Find $\sin (-240°)$.

Solution. $\sin (-240°) = -\sin 240°$
$$= -(-\sin 60°) = \tfrac{1}{2}\sqrt{3}.$$

(2) Find tan $(-195° 30')$.

Solution. tan $(-195° 30') = -\tan 195° 30'$
$$= -\tan 15° 30' = -.2773.$$

26. Variation of the Trigonometric Functions. Let us look upon the trigonometric functions as special functions of an independent variable (the angle) as we defined functions in Section 16. Thus we can write, for example:

$$y = f(\theta) = \sin \theta.$$

If we let θ vary arbitrarily we can calculate a table of values of y:

θ	0°	30°	60°	90°	120°	150°	180°	210°	240°	270°	300°	330°	360°
$y = \sin \theta$	0	.5	.87	1	.87	.5	0	−.5	−.87	−1	−.87	−.5	0

Let us study this table closely. We notice that as θ varies from 0° to 90°, sin θ varies from 0 to 1; then for $90° \leq \theta \leq 180°$, we have:

$$1 \geq \sin \theta \geq 0; \quad 180° \leq \theta \leq 270°, \quad 0 \geq \sin \theta \geq -1;$$

and

$$270° \leq \theta \leq 360°, \quad -1 \leq \sin \theta \leq 0.$$

Thus, as θ increases steadily we notice that sin θ increases for a while, then decreases, and after a while increases again. The precise variation can be summarized in a table.

As θ increases from	0° to 90°	90° to 180°	180° to 270°	270° to 360°
sin θ	Inc. 0 to 1	Dec. 1 to 0	Dec. 0 to −1	Inc. −1 to 0

In Section 22 we defined the trigonometric functions for all angles, with the exception of certain quadrantal angles. For example, the tangent of 90° was left undefined since we cannot divide by zero. To study the complete variation of the trigonometric functions we shall show that they can be associated with the length of a line and, in fact, can be defined in terms of a directed line segment. Place a circle of radius 1 on a rectangular coordinate plane so that its center is at the origin, $\mathcal{O}$. Place an angle θ in the standard position and let P be the point of intersection of the terminal side and the circle. Let N be the foot of the perpendicular from P to the X-axis. See Fig. 14(a).

In Fig. 14(b) we choose P on the terminal side of θ and such that PN is tangent to the circle at N.

Now from the definitions in Section 22 we have:

$$\sin \theta = \frac{y}{r} = \frac{NP}{1} = NP; \quad \cos \theta = \frac{x}{r} = \frac{ON}{1} = ON;$$

$$\sec \theta = \frac{r}{x} = \frac{1}{ON}; \quad \csc \theta = \frac{r}{y} = \frac{1}{NP};$$

end in Fig. 14(b):

$$\tan \theta = \frac{y}{x} = \frac{NP}{1} = NP \qquad \text{for} \quad 0 \le \theta < 90°;$$

$$\tan \theta = \frac{y}{x} = \frac{N'P'}{-1} = -N'P' \quad \text{for} \quad 90° < \theta \le 180°.$$

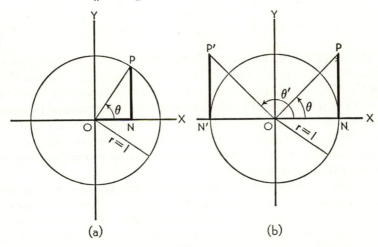

(a)   (b)

Fɪɢ. 14

The directed line segment NP in Fig. 14(a) varies in value from 0 to 1 to 0 to -1 to 0 as the angle moves through one revolution, which agrees with the variation for $\sin \theta$ given above. The directed line segment NP in Fig. 14(b) steadily increases in size as θ varies from 0 to $\frac{\pi}{2}$ and finally becomes larger than any number we have. As soon as θ passes $\frac{\pi}{2}$ the tangent becomes a very large negative number and then increases to zero as θ approaches and becomes π. Even though there is no value for $\tan \theta$ at $\theta = \frac{\pi}{2}$, we express its behavior near $\frac{\pi}{2}$ symbolically by:

$$\tan\left(\frac{\pi}{2}\right)^{-} = +\infty \quad \text{and} \quad \tan\left(\frac{\pi}{2}\right)^{+} = -\infty,$$

where ∞ is the symbol called *infinity*.

With this symbolic concept we can give the variation of six trigonometric functions in summarized form.

θ	0° to 90°	90° to 180°	180° to 270°	270° to 360°
sin θ	0 to 1	1 to 0	0 to -1	-1 to 0
cos θ	1 to 0	0 to -1	-1 to 0	0 to 1
tan θ	0 to $+\infty$	$-\infty$ to 0	0 to $+\infty$	$-\infty$ to 0
cot θ	$+\infty$ to 0	0 to $-\infty$	$+\infty$ to 0	0 to $-\infty$
sec θ	1 to $+\infty$	$-\infty$ to -1	-1 to $-\infty$	$+\infty$ to 1
csc θ	$+\infty$ to 1	1 to $+\infty$	$-\infty$ to -1	-1 to $-\infty$

27. Graphs of the Trigonometric Functions. If we let θ be replaced by x, then we have $y = \sin \theta = \sin x$, and we may graph this function on the (x,y)-plane. That is, we can consider corresponding values of x and y as coordinates of a point and plot it on a system of rectangular coordinates, thus obtaining the graphs of the trigonometric functions. For this kind of plotting we shall keep the same units of length on the x and y axes and then it is necessary to first change x to radian measure in order to get a true picture.

To plot the graphs of the trigonometric functions first change the angle to radian measure, then calculate the values of the trigonometric functions, and finally plot corresponding values. The table of values need only be recorded once. It is given on p. 41 at intervals of 30° or $\pi/6$, and the values are correct to two decimal places. The extreme right column gives the angle in radians correct to two decimal places. The graphs of the trigonometric functions are shown in Figs. 15–20.

28. Inverse Trigonometric Functions. In order to discuss the inverse trigonometric functions we need some new terminology.

Definition. *The expression* **arcsin x** *means an* **angle whose sine is x.** This leads to two equivalent equations:

$$y = \text{arcsin } x \quad \text{and} \quad x = \sin y,$$

which first express y as a function of x and then express x as a function of y. Thus we say that arcsin x and sin y are **inverse functions.** We cannot stress too strongly that arcsin x is a notation for an **angle** or a pure number which may be associated with an angle.

Note: The angle arcsin x is also written $\sin^{-1} x$. This is sometimes

x	$\sin x$	$\cos x$	$\tan x$	$\cot x$	$\sec x$	$\csc x$	x
0	0	1.00	0	$\pm\infty$	1.00	$\pm\infty$	0
$\dfrac{\pi}{6}$	0.50	0.87	0.58	1.73	1.15	2.00	0.52
$\dfrac{\pi}{3}$	0.87	0.50	1.73	0.58	2.00	1.15	1.05
$\dfrac{\pi}{2}$	1.00	0	$\pm\infty$	0	$\pm\infty$	1.00	1.57
$\dfrac{2\pi}{3}$	0.87	-0.50	-1.73	-0.58	-2.00	1.15	2.09
$\dfrac{5\pi}{6}$	0.50	-0.87	-0.58	-1.73	-1.15	2.00	2.62
π	0	-1.00	0	$\pm\infty$	-1.00	$\pm\infty$	3.14
$\dfrac{7\pi}{6}$	-0.50	-0.87	0.58	1.73	-1.15	-2.00	3.67
$\dfrac{4\pi}{3}$	-0.87	-0.50	1.73	0.58	-2.00	-1.15	4.19
$\dfrac{3\pi}{2}$	-1.00	0	$\pm\infty$	0	$\pm\infty$	-1.00	4.71
$\dfrac{5\pi}{3}$	-0.87	0.50	-1.73	-0.58	2.00	-1.15	5.24
$\dfrac{11\pi}{6}$	-0.50	0.87	-0.58	-1.73	1.15	-2.00	5.76
2π	0	1.00	0	$\pm\infty$	1.00	$\pm\infty$	6.28

confused with the negative exponent notation, and the student is cautioned against this common error; i.e., $\sin^{-1} x \neq 1/\sin x$.

Illustrations.

(1) Find $\arcsin \frac{1}{2}$.

Solution. Let $y = \arcsin \frac{1}{2}$; then $\sin y = \frac{1}{2}$. The angle corresponding to this function is 30°, 150°. However, the sine function has a period of 2π, and any angle obtained by adding $n(2\pi)$ to 30° or 150° would satisfy the equation. Thus:

$$\arcsin \tfrac{1}{2} = \begin{cases} 2n\pi + \tfrac{1}{6}\pi \\ 2n\pi + \tfrac{5}{6}\pi \end{cases} \quad \text{or} \quad \begin{cases} n(360°) + 30° \\ n(360°) + 150°, \end{cases}$$

where n may be any integer.

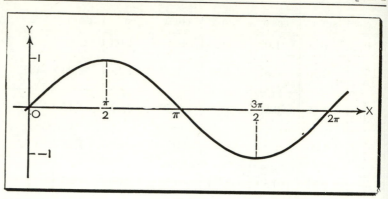

Fig. 15

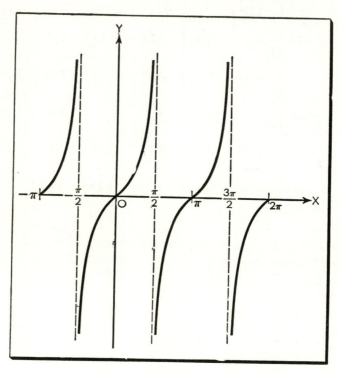

Fig. 17

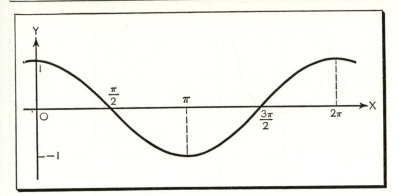

Fɪɢ. 16

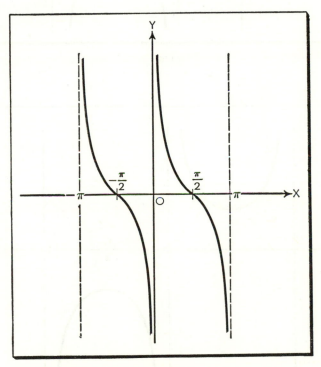

Fɪɢ. 18

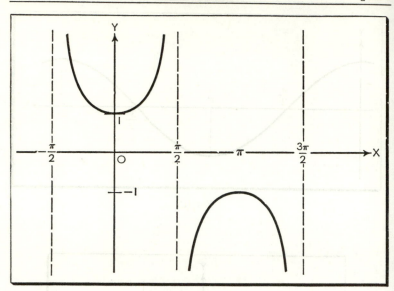

FIG. 19

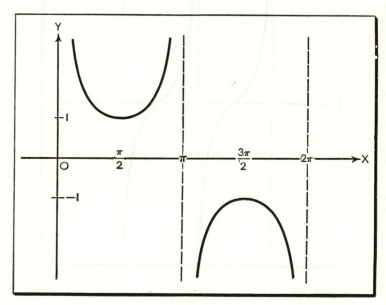

FIG. 20

(2) Find $\sin \arccos \frac{12}{13}$ for a first-quadrant angle.

Solution. Let $\theta = \arccos \frac{12}{13}$; then $\cos \theta = \frac{12}{13}$. Construct the triangle of Fig. 21 and from it we obtain $\sin \theta = \frac{5}{13}$. Therefore:

$$\sin \arccos \tfrac{12}{13} = \tfrac{5}{13}.$$

If we want to consider only the smallest numerical value of the angle, we give it a special name.*

Definition. *The* **principal value** *of arcsin x, arccsc x, arctan x, and arccot x is the* **smallest numerical** *value; of the arccos x and arcsec x it is the* **smallest positive** *value of the angle.*

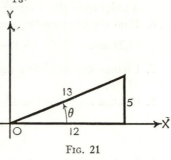

Fig. 21

The principal value is usually indicated by capitalizing the "A" in "arc." These values can be summarized in a table:

$$-\frac{\pi}{2} \leq \text{Arcsin } x \leq \frac{\pi}{2}; \quad -\frac{\pi}{2} \leq \text{Arccot } x \leq \frac{\pi}{2};$$

$$0 \leq \text{Arccos } x \leq \pi; \quad 0 \leq \text{Arcsec } x \leq \pi;$$

$$-\frac{\pi}{2} \leq \text{Arctan } x \leq \frac{\pi}{2}; \quad -\frac{\pi}{2} \leq \text{Arccsc } x \leq \frac{\pi}{2}.$$

Illustrations.
(1) $\text{Arccot } (-\sqrt{3}) = -30°$.
(2) $\text{Arcsin } .6428 = 40°$.
(3) $\text{Arctan } (-1) = -45°$.
(4) $\text{Arcsec } (-\sqrt{2}) = 135°$.
(5) $\text{Arccos } .5422 = 57° 10'$.

We notice that the values of x in arcsin x and arccos x are limited to the interval $-1 \leq x \leq 1$. The graphs of the inverse functions are the same as the graphs of the trigonometric functions with an interchange of the X- and Y-axes.

29. Exercise II.

1. Find the values of the function $y = 9 - x^2$ for $x = -4, -3, -2, -1, 0, 1, 2, 3, 4$. Find the values of Δy and $\Delta y/\Delta x$.

---

* It is regrettable that we do not have uniformity in the definition of principal values. It is believed that the definition given here agrees with a majority of the existing elementary books. The student, however, is warned that others exist.

2. Graph the function of Problem 1.
3. Find the zeros and intercepts of Problem 1.
4. Plot the function $y = x^3 - 6x^2 + 11x - 6$. Find its intercepts.
5. What is the locus of a point which moves so that it is always 3 units from the Y-axis?
6. Find the distance between the following points:

   (3,2) and (−1,0); (−3,−2) and (0,3); (8,3) and (4,1).

7. Change the following angles to radian measure: 30°, 90°, 5°, 3° 10′.
8. Change the following angles to sexagesimal measure:

$$\frac{\pi}{4}, \frac{5\pi}{6}, 3 \text{ radians.}$$

9. Find each function by use of a table:

   (a) sin 68° 10′. (d) cot 73° 22′. (g) sin 13° 52′.
   (b) cos 33° 15′. (e) cos 61° 14′. (h) cot 29° 40′.
   (c) tan 37° 18′. (f) tan 78° 20′. (i) sin 45° 33′.

10. Find the acute angle α.

   (a) sin α = .5324. (d) tan α = 1.1193. (g) cot α = .9380.
   (b) cos α = .6734. (e) cos α = .5885. (h) cos α = .7363.
   (c) tan α = .8847. (f) sin α = .7284. (i) sin α = 1.3291.

11. Find the values of the following functions:

   (a) sin 432°. (d) cot (−18° 31′).
   (b) cos 167°. (e) sin (−156° 12′).
   (c) tan 231° 10′. (f) cos (−243° 40′).

12. Plot the graph of sin $3x$.
13. Find sin arccos $\frac{3}{5}$.
14. Find Arcsin .1363.
15. Calculate a table of values for $y = 2 \sin 2x$; $0 \leq x \leq 6.28$. Plot the function and find the distance between the points at

   $x = \dfrac{\pi}{4}$ and $x = \dfrac{5\pi}{4}$.

3

EQUATIONS AND IDENTITIES

30. Definitions. The student is already familiar with the equal sign and the basic concepts of equations. Nevertheless, let us define our terms for the discussion which follows.

An **equation** *is a statement that two expressions are equal.* The expressions are called **members** or **sides** of the equality, and we speak of them as the *left member* or *right member* and as the *left-hand side* or *right-hand side*.

The equality may be true at all times and for all values of the symbols for which the members are defined, and it is then called an **identity.**

Illustration.

$$x^2 - y^2 = (x - y)(x + y).$$

An equality which is true for only certain values of the symbols is called a **conditional equality** or more often simply an **equation.**

Illustration.

$$x + 3 = 7 \text{ is true only if } x = 4.$$

The equal sign, $=$, is used for both equations and identities; however, the symbol $\equiv$, "is identically equal to," is sometimes used to emphasize an identity. In an equation the symbol which is not known is logically called an **unknown.** Quantities which are known are referred to as *constants* and agree with our definition in the last chapter. There are many kinds of equations and we shall have occasion to meet a large class of them in this book.

Our main interest in equations is usually to find the value of the unknown which makes it a true statement. This value is called a solution of the equation. More precisely, *a value of the unknown which when substituted into the equation makes its two members equal is called a* **solution.** Each such solution is called a **root** of the equation.

We may change the form of an equation, but in so doing we must

47

be careful not to destroy its original meaning. If we change an equation to another form we have a derived equation, and not to lose the
intent of the original equation we say that:

A derived equation is **equivalent** *to the original equation if it contains
all the roots of that equation and no more.*

There are two main operations which lead to equivalent equations:

1. *We may add the same number or expression to, or subtract the same
number or expression from, both members of the equation.*

2. *We may multiply or divide both members of the equation by the
same number or expression, providing it is not zero or does not contain
the unknown.*

Illustrations.

(1) $2x = 5$ and $2x + 3 = 8$ are equivalent.

(2) $2x = 5$ and $4x = 10$ are equivalent.

The first of these operations leads to the well-known concept of **transposing** terms from one side to the other, since this is simply the operation of subtracting the same quantity from both sides.

Illustration. If we transpose 2 from the left member to the right member
in the equation $3x + 2 = 5$ we obtain $3x = 5 - 2 = 3$, which can also be
obtained by subtracting 2 from both sides.

Although the multiplication of both members by an expression
containing the unknown does not lead to equivalent equations, it is
sometimes done and may result in an equation which has roots that
are *not* roots of the original equation. Such roots are called **extraneous
roots,** and the derived equation is said to be **redundant** with respect
to the original equation if it contains all the roots of that equation plus
some others.

We may also divide both members of an equation by an expression
which contains the unknown, but in so doing we may lose some roots.
The derived equation is said to be **defective** with respect to the original
equation if it does not have all the roots of that equation.

Illustrations.

(1) Consider the equation $x + 2 = 3$ which has the root $x = 1$. Multiply
both members by $x + 1$; the result is $x^2 + 3x + 2 = 3x + 3$ or $x^2 = 1$,
which has the roots $x = \pm 1$. The root -1, however, is not a root of
the original equation.

(2) Consider the equation $(x - 2)(x + 2) = 0$, which has roots $x = 2$

and $x = -2$. Divide both members by $x - 2$ to get $x + 2 = 0$, which has the root $x = -2$ but does not have the other root.

One of the first equations we meet is the **polynomial equation** in one unknown:

$$a_0x^n + a_1x^{n-1} + a_2x^{n-2} + \cdots + a_{n-1}x + a_n = 0,$$

where n is a positive integer and the coefficients $a_0, a_1, a_2, \ldots, a_n$ are constants. The **degree** of this equation is the degree of the term that is highest. We give special names to some of the lower-degree equations:

Degree	Name	Equation
first	linear	$ax + b = 0$
second	quadratic	$ax^2 + bx + c = 0$
third	cubic	$ax^3 + bx^2 + cx + d = 0$
fourth	quartic	$ax^4 + bx^3 + cx^2 + dx + e = 0$

31. Solving Equations. To solve an equation we seek to find its roots, and having found them we check by substituting the values into the original equation. To find the roots we perform operations which in so far as is possible lead to equivalent equations and arrive at a clear indication of the value of the root. For linear equations or equations reducible to linear equations this is fairly simple.

Illustration. Solve for x in $3x - 7 = 2x - 5$.
Solution.

　Subtract $2x$, yields: $x - 7 = -5$.
　Add　　7, yields:　　$x = 2$.
　Check: $3(2) - 7 = 2(2) - 5$　or　$-1 = -1$.

If the equation is given in fractional form, multiply through by the lowest common denominator (L.C.D.).

Illustration. Solve:

$$\frac{x}{x - 2} - \frac{2x}{x + 2} = 2 - \frac{3x^2}{x^2 - 4}.$$

Solution. Clear the equation of fractions by multiplying by the L.C.D. $(x - 2)(x + 2)$ and simplify:

$$x(x + 2) - 2x(x - 2) = 2(x^2 - 4) - 3x^2.$$
$$x^2 + 2x - 2x^2 + 4x = 2x^2 - 8 - 3x^2.$$
$$-x^2 + 6x = -x^2 - 8.$$
$$x = -\tfrac{4}{3}.$$

Check.

$$\frac{-\frac{4}{3}}{(-\frac{4}{3}) - 2} - \frac{2(-\frac{4}{3})}{(-\frac{4}{3}) + 2} = 2 - \frac{3(-\frac{4}{3})^2}{(-\frac{4}{3})^2 - 4}.$$

$$\tfrac{22}{5} = \tfrac{22}{5}.$$

In solving physical problems (to students, "word problems") the essential procedure is to convert the words in the problem into the symbols and equations of algebra. A thorough reading and understanding of the problem is the first step; the algebra itself is usually not very difficult.

Illustration. A man sold 2 acres less than $\frac{3}{4}$ of his farm. He then had 4 acres less than $\frac{1}{2}$ of it left. How many acres were on his original farm? *Solution.* Let x = number of acres in original farm. Using the principle that the sum of the parts equals the whole, we write:

$$\tfrac{3}{4}x - 2 + \tfrac{1}{2}x - 4 = x.$$

Solving for x we get:

$$(\tfrac{3}{4} + \tfrac{1}{2} - 1)x = 6.$$
$$x = 24.$$

Some concepts of the physical laws or assumptions must be known before word problems can be translated into algebraic equations. We must, for example, know that the distance traveled equals velocity times time; that the area of a rectangle is the length times the width; etc. Many of these concepts will be explained in this book; others the student will already know; still others may be found in advanced books to which we may refer.

Illustration. A and B start from the same place and travel in the same direction with B starting $\frac{1}{2}$ hour after A. If A travels at a rate of 50 miles per hour, and B, 60 miles per hour, how far will they have traveled when B overtakes A? *Solution.* Let $t \equiv$ the number of hours B has traveled. Then $t + \frac{1}{2}$ $\equiv$ number of hours A has traveled. The distance, d, is the average speed times the time. Thus:

A travels: $d = 50(t + \tfrac{1}{2})$.
B travels: $d = 60t$.

Now B overtakes A when they have both traveled the same distance, so that:

$$60t = 50(t + \tfrac{1}{2}).$$

Solving for t, we have:

$$10t = 25 \quad \text{or} \quad t = 2.5 \text{ hours.}$$

Check. As a check let us compute the distance traveled.

For A: $d = 50(t + \frac{1}{2}) = 50(3) = 150$ miles.
For B: $d = 60t = 60(2.5) = 150$ miles.

A very important principle in the solution of equations is the one: *If a* **product** *of two or more factors equals zero, one or more of its* **factors** *equals zero.*

Illustration. In solving the equation:

$$(3x - 9)(x + 2) = 0$$

we set each factor equal to zero.

$3x - 9 = 0.$	$x + 2 = 0.$
$x = 3.$	$x = -2.$

Check. $\qquad (3 \cdot 3 - 9)(3 + 2) = (0)(5) = 0.$
$[(3)(-2) - 9][-2 + 2] = (-15)(0) = 0.$

32. Fundamental Trigonometric Identities. Although identities are important in algebra (see for example the algebraic identities in Section 8), they are equally important in trigonometry. We shall begin with the **fundamental identities,** of which there are eight:

The reciprocal relations.

1. $\csc \theta = \dfrac{1}{\sin \theta}$; **2.** $\sec \theta = \dfrac{1}{\cos \theta}$; **3.** $\cot \theta = \dfrac{1}{\tan \theta}.$

The quotient relations.

4. $\tan \theta = \dfrac{\sin \theta}{\cos \theta}$; **5.** $\cot \theta = \dfrac{\cos \theta}{\sin \theta}.$

The Pythagorean relations.

6. $\sin^2 \theta + \cos^2 \theta = 1;$
7. $\tan^2 \theta + 1 = \sec^2 \theta;$
8. $1 + \cot^2 \theta = \csc^2 \theta.$

To prove these identities we return to the definitions of Section 22.

Proof of (1):

By the definition we have $\csc \theta = \dfrac{r}{y}$ and $\sin \theta = \dfrac{y}{r}.$

$\therefore \quad \csc \theta = \dfrac{r}{y} = \dfrac{1}{\dfrac{y}{r}} = \dfrac{1}{\sin \theta}.$

Proof of (4):

By the definition we have $\sin \theta = \dfrac{y}{r}$, $\cos \theta = \dfrac{x}{r}$, $\tan \theta = \dfrac{y}{x}$.

$$\therefore \quad \frac{\sin \theta}{\cos \theta} = \frac{\dfrac{y}{r}}{\dfrac{x}{r}} = \frac{y}{r} \cdot \frac{r}{x} = \frac{y}{x} = \tan \theta.$$

Proof of (7):

By the Pythagorean theorem we have $y^2 + x^2 = r^2$.

Divide through by x^2 to get $\dfrac{y^2}{x^2} + 1 = \dfrac{r^2}{x^2}$.

Then, from the definitions, $\tan^2 \theta + 1 = \sec^2 \theta$.

The student should prove all the identities and then commit them to memory.

The fundamental identities may take different forms. For example, since:

$$\csc \theta = \frac{1}{\sin \theta}, \quad \text{we have} \quad \sin \theta = \frac{1}{\csc \theta}.$$

Similarly, from:

$$\sin^2 \theta + \cos^2 \theta = 1, \quad \text{we have} \quad 1 - \sin^2 \theta = \cos^2 \theta.$$

The fundamental identities are used to prove other trigonometric identities, of which there are a large number.

Illustrations.

(1) Prove the identity $\tan \alpha + \cot \alpha = \sec \alpha \csc \alpha$.

Solution. We shall alter only the left member.

Using (4) and (5): $\qquad \dfrac{\sin \alpha}{\cos \alpha} + \dfrac{\cos \alpha}{\sin \alpha} =$

Simplifying: $\qquad \dfrac{\sin^2 \alpha + \cos^2 \alpha}{\cos \alpha \sin \alpha} =$

Using (6): $\qquad \dfrac{1}{\cos \alpha \sin \alpha} =$

or $\qquad \dfrac{1}{\cos \alpha} \dfrac{1}{\sin \alpha} =$

Using variations of (1) and (2): $\quad \dfrac{1}{\cos \alpha} \dfrac{1}{\sin \alpha} = \sec \alpha \csc \alpha.$

(2) Prove the identity $\cos \alpha = \dfrac{\csc \alpha}{\cot \alpha + \tan \alpha}$.

Solution.　Change the right member.

By (1), (4), and (5):　$\cos \alpha = \dfrac{\dfrac{1}{\sin \alpha}}{\dfrac{\cos \alpha}{\sin \alpha} + \dfrac{\sin \alpha}{\cos \alpha}}$

Simplifying:　　　　　　　$= \dfrac{\dfrac{1}{\sin \alpha}}{\dfrac{\cos^2 \alpha + \sin^2 \alpha}{\sin \alpha \cos \alpha}}$

　　　　　　　　　　　　$= \dfrac{1}{\sin \alpha} \cdot \dfrac{\sin \alpha \cos \alpha}{\cos^2 \alpha + \sin^2 \alpha}$

By (6):　　　　　　　　$= \cos \alpha.$

Here are some suggestions for proving trigonometric identities:

1. All functions may be easily expressed in terms of sines and cosines.
2. If one member involves only one function, express everything on the other side in terms of this function.

33. Trigonometric Equations.　In solving trigonometric equations we realize that because the trigonometric functions are periodic, any such equation has infinitely many solutions.　Consequently, in solving these equations we are primarily concerned with finding the angles which are positive or zero and less than 360°; i.e., $0 \le \theta < 360°$. The method of solution can best be explained by considering some examples.

Illustrations.

(1) Solve: $2 \cos \theta - 1 = 0$.

Solution. $2 \cos \theta - 1 = 0$ yields $\cos \theta = \frac{1}{2}$.

We recall that $\cos 60° = \frac{1}{2}$ and that the cosine is positive in the first and fourth quadrants; therefore $\theta = 60°$ and $300°$.

(2) Solve: $2 \cos \theta + 1.9836 = 0$.

Solution.　Transposing and dividing yields:

$$\cos \theta = -.9918.$$

By use of the table we have a reference angle of 7° 20′.　The cosine is negative in the quadrants II and III.　Thus:

$$\theta = 180° - 7° 20' = 172° 40' \quad \text{and} \quad \theta = 180° + 7° 20' = 187° 20'.$$

(3) Solve: $2 \sin^2 \theta - \sin \theta - 1 = 0$.

Solution. This is of the form $2y^2 - y - 1$, and by the factoring formulas of Section 8 we may write that as $(2y + 1)(y - 1)$. Thus we have:

$2 \sin \theta + 1 = 0.$	$\sin \theta - 1 = 0.$
$\sin \theta = -\frac{1}{2}.$	$\sin \theta = 1.$
$\theta = 210°, 330°.$	$\theta = 90°.$

(4) Solve $\sqrt{3} \cot \theta + 1 = \csc^2 \theta$.

Solution. First change the equation by use of identities.

$$\sqrt{3} \cot \theta + 1 = 1 + \cot^2 \theta.$$

Then: $$\cot^2 \theta - \sqrt{3} \cot \theta = 0.$$
$$\cot \theta \, (\cot \theta - \sqrt{3}) = 0.$$

$\cot \theta = 0.$	$\cot \theta = \sqrt{3},$
$\theta = 90°, 270°.$	$\theta = 30°, 210°.$

34. Addition Formulas. A large portion of the analytical part of trigonometry is the process of changing from one expression to another. We have already studied the fundamental identities, and we shall now turn to a set of formulas known as the addition formulas.

I. The sum of two angles.

Let α and β be any two angles; then:

1. $\sin (\alpha + \beta) = \sin \alpha \cos \beta + \cos \alpha \sin \beta$;

2. $\cos (\alpha + \beta) = \cos \alpha \cos \beta - \sin \alpha \sin \beta$;

3. $\tan (\alpha + \beta) = \dfrac{\tan \alpha + \tan \beta}{1 - \tan \alpha \tan \beta}$;

4. $\cot (\alpha + \beta) = \dfrac{\cot \alpha \cot \beta - 1}{\cot \alpha + \cot \beta}$.

Proof. To prove these formulas it is advantageous to use a geometric figure for the first two; the remaining ones can then be proved by the use of identities. Let us prove formulas (1) and (2) for the case when α and β are positive acute angles. Construct Fig. 22.

The construction is such that $PQ \perp OP$ and $QN \perp OX$; therefore by elementary geometry $\angle RQP = \alpha$. The triangles ONQ, OMP, and PRQ are right triangles. Using the right triangle definition, we have:

$$\triangle OMP: \quad \sin \alpha = \frac{PM}{OP}; \quad PM = OP \sin \alpha; \tag{i}$$

$$\cos \alpha = \frac{OM}{OP}; \quad OM = OP \cos \alpha; \tag{ii}$$

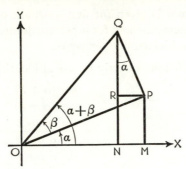

Fɪɢ. 22

$\triangle PRQ:$ $\sin \alpha = \dfrac{PR}{QP};$ $PR = QP \sin \alpha;$ (iii)

$\cos \alpha = \dfrac{QR}{QP};$ $QR = QP \cos \alpha;$ (iv)

$\triangle OPQ:$ $\sin \beta = \dfrac{QP}{OQ};$ $\cos \beta = \dfrac{OP}{OQ}.$ (v)

Now $RN = PM$, so that in $\triangle ONQ$ we have:

$$NQ = QR + RN = QR + PM,$$

so that:

$$\sin (\alpha + \beta) = \frac{NQ}{OQ} = \frac{PM}{OQ} + \frac{QR}{OQ}.$$

Using (i) and (iv) we get:

$$\sin (\alpha + \beta) = \frac{OP}{OQ} \sin \alpha + \frac{QP}{OQ} \cos \alpha;$$

and with (v) we obtain:

$$\sin (\alpha + \beta) = \sin \alpha \cos \beta + \cos \alpha \sin \beta.$$

Furthermore:

$$\cos (\alpha + \beta) = \frac{ON}{OQ} = \frac{OM}{OQ} - \frac{NM}{OQ},$$

and we constructed the figure so that $NM = PR$. Using the expressions (ii), (iii), and (v), we easily see that:

$$\cos (\alpha + \beta) = \frac{OP}{OQ} \cos \alpha - \frac{QP}{OQ} \sin \alpha$$

$$= \cos \alpha \cos \beta - \sin \alpha \sin \beta.$$

Although the proof is given for first-quadrant angles, it is quite general, and all statements hold for other quadrants provided careful attention is paid to directed line segments.

Proof of (3). This follows directly by use of identities.

$$\tan (\alpha + \beta) = \frac{\sin (\alpha + \beta)}{\cos (\alpha + \beta)}$$

[By (1) and (2)]:
$$= \frac{\sin \alpha \cos \beta + \cos \alpha \sin \beta}{\cos \alpha \cos \beta - \sin \alpha \sin \beta}$$

[Divide by $\cos \alpha \cos \beta$]:
$$= \frac{\dfrac{\sin \alpha \cos \beta}{\cos \alpha \cos \beta} + \dfrac{\cos \alpha \sin \beta}{\cos \alpha \cos \beta}}{\dfrac{\cos \alpha \cos \beta}{\cos \alpha \cos \beta} - \dfrac{\sin \alpha \sin \beta}{\cos \alpha \cos \beta}}$$

[Simplify]:
$$= \frac{\dfrac{\sin \alpha}{\cos \alpha} + \dfrac{\sin \beta}{\cos \beta}}{1 - \dfrac{\sin \alpha}{\cos \alpha} \dfrac{\sin \beta}{\cos \beta}}$$

[By Fund. Identity]:
$$= \frac{\tan \alpha + \tan \beta}{1 - \tan \alpha \tan \beta}.$$

II. The difference of two angles.

5. $\sin (\alpha - \beta) = \sin \alpha \cos \beta - \cos \alpha \sin \beta$;

6. $\cos (\alpha - \beta) = \cos \alpha \cos \beta + \sin \alpha \sin \beta$;

7. $\tan (\alpha - \beta) = \dfrac{\tan \alpha - \tan \beta}{1 + \tan \alpha \tan \beta}$;

8. $\cot (\alpha - \beta) = \dfrac{\cot \alpha \cot \beta + 1}{\cot \beta - \cot \alpha}$.

Proof. These formulas are easily proved by simply substituting $(-\beta)$ for β in formulas (1), (2), (3), and (4) and recalling the definitions of the trigonometric functions of the negative angles (see Section 25). For example:

$$\sin (\alpha - \beta) = \sin (\alpha + [-\beta])$$
$$= \sin \alpha \cos (-\beta) + \sin (-\beta) \cos \alpha$$
$$= \sin \alpha \cos \beta - \sin \beta \cos \alpha.$$

III. Double-Angle Formulas.

9. $\sin 2\alpha = 2 \sin \alpha \cos \alpha$;

10. $\cos 2\alpha = \cos^2 \alpha - \sin^2 \alpha$
$$= 2 \cos^2 \alpha - 1$$
$$= 1 - 2 \sin^2 \alpha;$$

$$11. \ \tan 2\alpha = \frac{2 \tan \alpha}{1 - \tan^2 \alpha};$$

$$12. \ \cot 2\alpha = \frac{\cot^2 \alpha - 1}{2 \cot \alpha}.$$

Proof. The proofs of these formulas are easily obtained by letting $\beta = \alpha$ in formulas (1), (2), (3), and (4). For example:

$$\cos 2\alpha = \cos (\alpha + \alpha) = \cos \alpha \cos \alpha - \sin \alpha \sin \alpha$$
$$= \cos^2 \alpha - \sin^2 \alpha.$$

IV. Half-Angle Formulas.*

$$13. \ \sin \frac{\theta}{2} = \pm \sqrt{\frac{1 - \cos \theta}{2}};$$

$$14. \ \cos \frac{\theta}{2} = \pm \sqrt{\frac{1 + \cos \theta}{2}};$$

$$15. \ \tan \frac{\theta}{2} = \pm \sqrt{\frac{1 - \cos \theta}{1 + \cos \theta}} = \frac{1 - \cos \theta}{\sin \theta} = \frac{\sin \theta}{1 + \cos \theta};$$

$$16. \ \cot \frac{\theta}{2} = \pm \sqrt{\frac{1 + \cos \theta}{1 - \cos \theta}} = \frac{\sin \theta}{1 - \cos \theta} = \frac{1 + \cos \theta}{\sin \theta}.$$

Proof. The proofs of the first two formulas are obtained from formula (10). First, use:

$$\cos 2\alpha = 1 - 2 \sin^2 \alpha$$

and solve for $\sin \alpha$:

$$\sin^2 \alpha = \tfrac{1}{2}(1 - \cos 2\alpha)$$

and

$$\sin \alpha = \pm \sqrt{\frac{1 - \cos 2\alpha}{2}}.$$

Let $\theta = 2\alpha$; then $\frac{\theta}{2} = \alpha$ and we get (13). Now use:

$$\cos 2\alpha = 2 \cos^2 \alpha - 1$$

and solve for $\cos \alpha$; the result is (14). To obtain (15) and (16) replace the tangent and cotangent by the equivalent formula in terms of sine and cosine; then use formulas (13) and (14) and simplify the

* The $\pm$ signs are somewhat ambiguous. Both or only one may apply. For a given θ they are determined from the quadrant in which $\frac{\theta}{2}$ falls.

radicals. This forms a good exercise in the simplification of radicals; the student should carry out each step and should rationalize both the numerator and the denominator. The $\pm$ sign is necessary before the radicals, but the student may wonder why it is not carried over after rationalization. It is not necessary because $1 - \cos \theta$ and $1 + \cos \theta$ are always positive and $\sin \theta$ will always have the same sign as $\tan \frac{1}{2}\theta$.

V. Product Formulas. Since all functions can be expressed in terms of sines and cosines, only the product formulas for these will be given.

> **17.** $2 \sin \alpha \cos \beta = \sin (\alpha + \beta) + \sin (\alpha - \beta)$;
> **18.** $2 \cos \alpha \cos \beta = \cos (\alpha + \beta) + \cos (\alpha - \beta)$;
> **19.** $2 \sin \alpha \sin \beta = \cos (\alpha - \beta) - \cos (\alpha + \beta)$;
> **20.** $2 \sin \beta \cos \alpha = \sin (\alpha + \beta) - \sin (\alpha - \beta)$.

Proof. The proofs of these formulas are obtained by adding and subtracting formulas (1), (2), (5), and (6). For example:

$$(1) \qquad \sin (\alpha + \beta) = \sin \alpha \cos \beta + \cos \alpha \sin \beta$$
$$(5) \qquad \underline{\sin (\alpha - \beta) = \sin \alpha \cos \beta - \cos \alpha \sin \beta}$$
$$\sin (\alpha + \beta) + \sin (\alpha - \beta) = 2 \sin \alpha \cos \beta$$

If we interchange α and β we see that formula (20) is equivalent to formula (17). It is suggested that the student prove this to his own satisfaction.

VI. Sums and Differences of Sines and Cosines.

> **21.** $\sin x + \sin y = 2 \sin \dfrac{x + y}{2} \cos \dfrac{x - y}{2}$;
>
> **22.** $\sin x - \sin y = 2 \cos \dfrac{x + y}{2} \sin \dfrac{x - y}{2}$;
>
> **23.** $\cos x + \cos y = 2 \cos \dfrac{x + y}{2} \cos \dfrac{x - y}{2}$;
>
> **24.** $\cos x - \cos y = -2 \sin \dfrac{x + y}{2} \sin \dfrac{x - y}{2}$.

Proof. The proofs of these formulas are obtained from formulas (17)–(20) by letting:

$$x = \alpha + \beta \quad \text{and} \quad y = \alpha - \beta$$

so that:

$$x + y = 2\alpha \quad \text{and} \quad x - y = 2\beta,$$

or

$$\alpha = \frac{x+y}{2} \quad \text{and} \quad \beta = \frac{x-y}{2}.$$

Now by direct substitution into (17), (18), (19), and (20) we obtain the above formulas.

This completes the classical addition formulas. To be thoroughly familiar with them, the student should commit them to memory, and if he has developed the proof for each of them this is not difficult. It may be noted that no addition formulas for secant and cosecant are given. Should these functions occur in any problem, they are handled by expressing them in terms of sine and cosine from the reciprocal identities and then simplifying.

Illustration. Prove the identity:

$$\sec \frac{\theta}{2} = \pm \sqrt{\frac{2}{1 + \cos \theta}}.$$

Solution. Use the fundamental identity and (14) to obtain:

$$\sec \frac{\theta}{2} = \frac{1}{\cos \dfrac{\theta}{2}} = \frac{1}{\pm \sqrt{\dfrac{1 + \cos \theta}{2}}}$$

$$= \pm \sqrt{\frac{2}{1 + \cos \theta}}.$$

35. Identities Involving Addition Formulas. The chief use of addition formulas is to change a given trigonometric expression. We shall illustrate by proving some identities and offer three suggestions:

1. Express secant and cosecant in terms of sine and cosine.

2. Reduce the multiple angles by changing to functions of single angles.

3. Avoid, as much as possible, the introduction of radicals.

Illustrations.

(1) Prove $\sin 4x = 8 \sin x \cos^3 x - 4 \sin x \cos x$.

Solution.
$$\begin{aligned}
\sin 4x &= \sin 2(2x) = 2 \sin 2x \cos 2x \\
&= 2[2 \sin x \cos x(2 \cos^2 x - 1)] \\
&= 4[2 \sin x \cos^3 x - \sin x \cos x] \\
&= 8 \sin x \cos^3 x - 4 \sin x \cos x.
\end{aligned}$$

(2) Prove $\tan (\tfrac{1}{2}x + 45°) - \tan x = \sec x$.

Solution.

$$\tan (\tfrac{1}{2}x + 45°) - \tan x = \frac{\tan \tfrac{1}{2}x + \tan 45°}{1 - \tan \tfrac{1}{2}x \tan 45°} - \tan x$$

$$= \frac{\tan \tfrac{1}{2}x + 1}{1 - \tan \tfrac{1}{2}x} - \tan x = \frac{\dfrac{\sin x}{1 + \cos x} + 1}{1 - \dfrac{\sin x}{1 + \cos x}} - \tan x$$

$$= \frac{\sin x + 1 + \cos x}{1 + \cos x - \sin x} - \frac{\sin x}{\cos x}$$

$$= \frac{\cos x + \cos^2 x - \sin x + \sin^2 x}{\cos x(1 + \cos x - \sin x)}$$

$$= \frac{(\cos x - \sin x + 1)}{\cos x(\cos x - \sin x + 1)}$$

$$= \frac{1}{\cos x} = \sec x.$$

(3) Prove $\dfrac{\cos 7x - \cos 5x}{\sin 5x - \sin 7x} = \tan 6x.$

Solution.

$$\frac{\cos 7x - \cos 5x}{\sin 5x - \sin 7x} = \frac{-2 \sin \dfrac{7x + 5x}{2} \sin \dfrac{7x - 5x}{2}}{2 \cos \dfrac{7x + 5x}{2} \sin \dfrac{5x - 7x}{2}}$$

$$= - \frac{\sin 6x \sin x}{\cos 6x \sin (-x)}$$

$$= \frac{\sin 6x}{\cos 6x} = \tan 6x.$$

36. Equations Involving Multiple Angles. Many trigonometry equations will involve multiple angles. To solve such an equation, we first reduce the equation to one or more equations each involving only one function of one angle and whose solutions include all solutions of the given equation.

Illustrations.

(1) Solve the equation $\sin 3x = \tfrac{1}{2}$.

Solution. Let $3x = \theta$; then we know that $\theta = 30° + n(360°)$ or $150° + n(360°)$. If we are interested only in positive values of x less than $360°$, we write:

$$\theta = 3x = 30°, 150°, 390°, 510°, 750°, 870°;$$
and $\qquad x = 10°, 50°, 130°, 170°, 250°, 290°.$

(2) Solve the equation $\cos 5x + \cos x = \cos 3x$.

Solution. $\cos 5x + \cos x = \cos 3x$.

By (23): $\qquad 2 \cos \dfrac{5x + x}{2} \cos \dfrac{5x - x}{2} = \cos 3x,$

or
$$2 \cos 3x \cos 2x - \cos 3x = 0.$$
$$\cos 3x (2 \cos 2x - 1) = 0.$$
$$\cos 3x = 0 \quad \text{and} \quad \cos 2x = \tfrac{1}{2}.$$

Thus:
$$2x = 60°, 300°, 420°, 660°;$$
$$3x = 90°, 270°, 450°, 630°, 810°, 990°;$$

and
$$x = 30°, 90°, 150°, 210°, 270°, 330°.$$

(3) Solve the equation $5 \sin 2x = 3$.

Solution.
$$\sin 2x = \tfrac{3}{5} = .6000.$$
$$2x = 36° 52'.$$
$$x = 18° 26'.$$

(4) Solve the equation $\cos 3x + \sin 2x - \sin 6x + \cos 5x = 0$.

Solution. By (23) and (22):
$$2 \cos 4x \cos x + 2 \cos 4x \sin (-2x) = 0.$$
$$2 \cos 4x [\cos x - \sin 2x] = 0.$$

$\cos 4x = 0.$	$\cos x - 2 \sin x \cos x = 0.$	
	$\cos x (1 - 2 \sin x) = 0.$	
$\cos 4x = 0;$	$\cos x = 0;$	$\sin x = \tfrac{1}{2};$
$4x = 90° + n360°;$	$x = 90° + n360°;$	$x = 30° + n360°;$
$4x = 270° + n360°;$	$x = 270° + n360°;$	$x = 150° + n360°;$

$$\therefore x = 30°, 90°, 150°, 270°, (22° 30' + n45°).$$

37. Summary. There are two kinds of equations: the conditional equation and the identity. Our concern with the conditional equation is usually that of finding the values of the unknown which satisfy the equation. To find these unknowns, or roots, we may change the original equation by algebraic manipulation involving the steps of arithmetic and known simplification rules. When the roots become obvious they should be checked in the original equation to establish the fact that they satisfy this equation. Frequently, there is more than one solution to an equation.

The identity is an expression which is true for all values of the symbols. Our chief concern with identities is to know of their existence, be certain that they are true, and use them to change given expressions to more desirable forms. There are many identities in the field of trigonometry, and the most useful formulas have been listed in this chapter. The student should commit most of these to memory and understand them thoroughly. Greater proficiency can be obtained by working many problems and it is recommended that all the exercises be solved.

38. Exercise III.

1. Solve the following equations:

   (a) $13x - 4 = 3x + 16$.

   (b) $\dfrac{3x - 9}{18} + \dfrac{x}{27} - \dfrac{2x - 5}{4} = \dfrac{4}{3} - x$.

   (c) $\dfrac{3}{x - 1} - \dfrac{2}{x + 4} = \dfrac{4}{2 - 2x}$.

   (d) $\dfrac{2x}{x - 3} - \dfrac{4x + 1}{2x - 1} = \dfrac{21}{2x^2 - 7x + 3}$.

   (e) $\dfrac{1}{4}(3y - 2) - \left[y - \dfrac{1}{y}(7 - 3y)\right] = -\dfrac{1}{4}y - 7$.

2. Prove the eight fundamental identities.

3. Prove the following trigonometric identities:

   (a) $\dfrac{\sin \alpha}{\csc \alpha} + \dfrac{\cos \alpha}{\sec \alpha} = 1$.

   (b) $\sin^4 y - \cos^4 y = 2 \sin^2 y - 1$.

   (c) $\dfrac{1}{1 - \sin \theta} + \dfrac{1}{1 + \sin \theta} = 2 \sec^2 \theta$.

   (d) $\dfrac{1 - \tan^2 x}{1 + \tan^2 x} = 1 - 2 \sin^2 x$.

   (e) $\sin \theta \sec \theta \cot \theta = 1$.

   (f) $\tan \theta \sin \theta + \cos \theta - \sec \theta = 0$.

4. Solve the following equations for values $< 360°$:

   (a) $\sin \theta = .3292$.

   (b) $2 \sin^2 x - 3 \cos x = 3$.

   (c) $\sin x \sec^2 x - 2 \sin x = 0$.

   (d) $\sin x + 1 = \cos x$. [Hint: square both sides.]

   (e) $\sec x \tan x = \tan x$.

5. Prove all the formulas (1)–(24) in Section 34.

6. Prove the following identities:

   (a) $\sin 3x = 3 \sin x - 4 \sin^3 x$.

   (b) $\dfrac{\sin 5x + \sin 3x}{\cos 5x + \cos 3x} = \tan 4x$.

   (c) $\cos^2 (45° - x) - \sin^2 (45° - x) = \sin 2x$.

   (d) $\dfrac{\cos x + 2 \cos 5x + \cos 9x}{\sin x + 2 \sin 5x + \sin 9x} = \cot 5x$.

(e) If $\alpha + \beta + \gamma = 180°$,

$$\tan \frac{\alpha}{2} \tan \frac{\beta}{2} + \tan \frac{\beta}{2} \tan \frac{\gamma}{2} + \tan \frac{\gamma}{2} \tan \frac{\alpha}{2} = 1.$$

7. Solve the following equations for values $< 360°$:
   (a) $\sin \frac{1}{2}x = \frac{1}{2}$.
   (b) $\cos 2x - \sin x = 0$.
   (c) $\sin 2x + \cos x = 0$.
   (d) $\dfrac{\tan x + \tan 2x}{1 - \tan x \tan 2x} = 1$.
   (e) $\sin 2\theta = \sin \theta$.

8. A rectangular picture is 2 times as long as it is wide and is sur-rounded by a frame 2 inches wide. Find its dimensions if the area of the framed picture is 64 square inches more than the area of the unframed picture.

9. Solve for f in the formula $D^2 = (f+s)/(f-r)$ and find the value if $D = 5$, $s = -2$, $r = 2$.

10. Find a number such that when 4 is subtracted from it, 2 plus $\frac{1}{5}$ of the remainder is equal to $\frac{1}{3}$ of the original number.

4

ALGEBRAIC TOOLS

39. Introduction. In the study of algebra there are certain concepts which not only aid us in understanding elementary algebra, but which also form the basis for certain phases of advanced mathematics. We shall call them algebraic tools and collect them together in this chapter. Some of the ideas may be new to the student; others may be familiar, depending upon his past study of high-school mathematics. These ideas should prove to be very interesting, for they find application not only in mathematics but also in other branches of the sciences and in logical thinking.

40. Ratio and Proportion. *A* **ratio** *is an indicated division.* It may be thought of as a fraction. The language used is "the ratio of a to b," which means $a \div b$ or a/b, and the symbol is $a:b$. In this notation a is the *first term* or the *antecedent,* and b is the *second term* or the *consequent.* It is important to remember that we treat the ratio as a fraction.

A **proportion** *is a statement that two ratios are equal.* Symbolically we write:

$$a:b = c:d \quad \text{or} \quad \frac{a}{b} = \frac{c}{d}.$$

The statement is read "a is to b as c is to d," and we call a and d the **extremes,** b and c the **means,** and d the **fourth proportional.** Proportions are treated as equations involving fractions. We may perform all the operations on them that we do on equations, and many of the resulting properties may already have been met in geometry.

Illustration. The volume in a container is proportional to the depth of the material in it. If the depth is 3 inches when the volume is 50 cubic feet, find the volume when the depth is 2 feet.

Solution. Let V = volume and d = depth. The ratio then states V/d, and the proportionality takes the form $V_1/d_1 = V_2/d_2$ where we have two different values of the volume and the depth. We now have given:

$$v_1 = 50 \text{ cu. ft.}; \quad d_1 = 3 \text{ in.} = \tfrac{1}{4} \text{ ft.}; \quad d_2 = 2 \text{ ft.}$$

so that:

$$\frac{50}{\tfrac{1}{4}} = \frac{V_2}{2} \quad \text{or} \quad 100 = \frac{1}{4} V_2;$$

thus:

$$V_2 = 400 \text{ cu. ft.}$$

Note that it is necessary to change to the same units of measurement.

41. Variation. There are three main kinds of variation.

I. Direct Variation. *If two variables, y and x are so related that the ratio y : x is always constant, then y* is said to vary directly as *x*.

We usually write this in the equation form:

$$\frac{y}{x} = k \quad \text{or} \quad y = kx,$$

and k is called the **proportionality factor** or the **constant of variation.**

Illustration. The area of a circle varies directly as the square of the radius. Find the constant of variation if the area is 16π when the radius is 4.

Solution. Let A = area and r = radius; then:

$$A = kr^2$$

and

$$16\pi = k(4)^2 = 16k$$

so that:

$$k = \pi.$$

II. Inverse Variation. *If two variables, x and y, are so related that y varies directly as the reciprocal of x, then y* is said to vary inversely as *x*.

We write this statement in the form:

$$y = \frac{k}{x}.$$

Illustration. The time it takes a car to travel a fixed distance varies inversely as its average speed. Find the average speed if it takes $2\tfrac{1}{2}$ hours to travel 95 miles.

Solution. Let v = average speed, t = time, and the constant distance $= d$. Then:

$$t = \frac{d}{v}$$

and

$$\frac{5}{2} = \frac{95}{v} \quad \text{or} \quad v = \frac{95}{5} \ (2) = 38.$$

III. Combined Variation. We are now in a position to expand this concept to more than two variables. However, in so doing it becomes necessary to add to our vocabulary.

If $z = kxy$, then z is said to **vary jointly** as x and y.

If $z = k\dfrac{x}{y}$, then z is said to *vary directly* as x and *inversely* as y.

Other combinations can be worded, and thus we call these **combined variations.**

Illustrations.

(1) Find the equation relating x, y, and z if y varies directly as the square of x and inversely as the cube of z, and if $y = 96$ when $x = \frac{9}{2}$ and $z = \frac{3}{4}$.

Solution. The equation is:

$$y = k\frac{x^2}{z^3}.$$

Substitution of the given values yields:

$$96 = k(\tfrac{9}{2})^2 \div (\tfrac{3}{4})^3 = k(\tfrac{81}{4})(\tfrac{64}{27}) = 48k,$$

or $$k = 2.$$

The relationship becomes:

$$y = 2\frac{x^2}{z^3}.$$

(2) The horsepower that a shaft can transmit varies jointly as its speed and the cube of its diameter. If a $2\frac{1}{4}$ inch shaft turning 125 r.p.m. transmits 110 horsepower, what horsepower can a 3 inch shaft transmit at 100 r.p.m.?

Solution.
$$\text{H.P.}_1 = kvd^3.$$
$$110 = k(125)(\tfrac{9}{4})^3.$$
$$k = \tfrac{22}{25}(\tfrac{4}{9})^3.$$
$$\text{H.P.}_2 = \tfrac{22}{25}(\tfrac{4}{9})^3(100)(3)^3$$
$$= 22(\tfrac{4}{3})^3(4)$$
$$= 208.6.$$

42. Binomial Theorem. The writing of a binomial to any integral power might necessitate long and tedious multiplications for large values of the exponent. To overcome this, mathematicians have proved a theorem which is most useful. The student can easily verify by multiplication that:

$$(a + b)^1 = a + b,$$
$$(a + b)^2 = a^2 + 2ab + b^2,$$
$$(a + b)^3 = a^3 + 3a^2b + 3ab^2 + b^3,$$
$$(a + b)^4 = a^4 + 4a^3b + 6a^2b^2 + 4ab^3 + b^4,$$
$$(a + b)^5 = a^5 + 5a^4b + 10a^3b^2 + 10a^2b^3 + 5ab^4 + b^5.$$

We seek an answer to the question, "What is the series of terms equal to $(a + b)^n$?" By noting the examples above for $n = 1, \ldots, 5$, we can check the following characteristics:

1. The first term is a^n.
2. The second term is $na^{n-1}b$.
3. The exponents of a decrease by **1** in succeeding terms, those of b increase by **1** in succeeding terms, and their sum in each term is **n**.
4. The coefficient of the "next" term is the coefficient of the preceding term multiplied by the exponent of a of that term and divided by the exponent of b of the "next" term.
5. There are $n + 1$ terms.
6. The coefficients of terms equidistant from the ends of the expansion are the same.

Assuming these characteristics to be true, we can write the **binomial formula**:

$$(a + b)^n = a^n + na^{n-1}b + \frac{n(n - 1)}{1 \cdot 2} a^{n-2}b^2 + \frac{n(n - 1)(n - 2)}{1 \cdot 2 \cdot 3} a^{n-3}b^3$$
$$+ \cdots + nab^{n-1} + b^n.$$

It may be desired to find only a particular term of this expansion without having to write out the entire expansion. We shall therefore try to develop a formula for a general term, and the term usually considered is the rth term. We can easily check the following features:

(i) The exponent of b is **1** less than the number of the term, that is, $r - 1$.

(ii) Since the sum of the exponents of a and b is n in each term, we have the exponent of a equal to $n - (r - 1)$.

(iii) The denominator of the coefficient is $1 \cdot 2 \cdot 3 \cdot \cdots (r - 1)$.

(iv) The numerator has the same number of factors as the denominator, thus is $n(n - 1)(n - 2) \cdots (n - r + 2)$.

The rth term then becomes:

$$\frac{n(n - 1)(n - 2) \cdots (n - r + 2)}{1 \cdot 2 \cdot 3 \cdot \cdots (r - 1)} a^{n-r+1}b^{r-1}.$$

Illustration. Expand $(x + 2y)^4$.

Solution.

$$(x + 2y)^4 = x^4 + 4x^3(2y) + \frac{4 \cdot 3}{1 \cdot 2} x^2(2y)^2$$

$$+ \frac{4 \cdot 3 \cdot 2}{1 \cdot 2 \cdot 3} x(2y)^3 + \frac{4 \cdot 3 \cdot 2 \cdot 1}{1 \cdot 2 \cdot 3 \cdot 4} (2y)^4$$

$$= x^4 + 8x^3y + 24x^2y^2 + 32xy^3 + 16y^4.$$

Illustration. Find the first three terms and the 7th term of the expansion of $(x^{\frac{1}{2}} - y^2)^{14}$.

Solution.

$$(x^{\frac{1}{2}} - y^2)^{14} = (x^{\frac{1}{2}})^{14} + 14(x^{\frac{1}{2}})^{13}(-y^2) + \frac{14 \cdot 13}{1 \cdot 2} (x^{\frac{1}{2}})^{12}(-y^2)^2 + \cdots$$

$$= x^7 - 14x^{\frac{13}{2}}y^2 + 91x^6y^4 - \cdots$$

The 7th term:

$$\frac{14 \cdot 13 \cdot 12 \cdot 11 \cdot 10 \cdot 9}{1 \cdot 2 \cdot 3 \cdot 4 \cdot 5 \cdot 6} (x^{\frac{1}{2}})^{14-7+1}(-y^2)^{7-1} = 3003x^4y^{12}.$$

43. Factorial Notation. The denominators of the coefficients of the terms in the binomial expansion have a particularly interesting product, and a special symbol is defined for such products.

The product of all the positive integers from **1** *to* **n**, *inclusive, is denoted by the symbol* **n!**:

$$1 \cdot 2 \cdot 3 \cdot 4 \cdot 5 \cdots n = n!$$

It is read "n factorial" or less commonly "factorial n." Another symbol used for the same notation is $\lfloor n$, but this is becoming obsolete. The values of factorials for $1 \leq n \leq 9$ are given in the following table.

n	$n!$
1	1
2	2
3	6
4	24
5	120
6	720
7	5040
8	40320
9	362880

An interesting property of factorials is that:

$$n! = n \cdot (n - 1)!$$

This permits us to *define* **0!**, for by letting $n = 1$ we have:

$$1! = 1 \cdot 0! \quad \text{or} \quad 0! = \frac{1}{1} = 1.$$

Another property is that if $r < n$, then:

$$\frac{n!}{r!} = n(n - 1) \cdots (r + 1).$$

44. Mathematical Induction. In our discussion of the binomial formula we noticed some characteristics of an expansion by studying a group of examples, wrote down a general formula, and verified it by more examples. This, of course, does not constitute a proof. There are many theorems in mathematics which have been discovered in this manner, and mathematicians have developed a method for proving the general case. The method is known as **mathematical induction,** and here is how it works. Suppose we know that a theorem is true for small integral values of n, say for $n = 1,2,3,\ldots,k$. We now state that *if it is true for $n = k$, then it is true for $n = k + 1$*, and proceed to prove this statement. It then follows that the theorem is true for all integral values of n. Consequently, the two important parts are: part I, the verification for $n = 1,2,3,\ldots,k$; and part II, the proof that if it is true for $n = k$, then it is true for $n = k + 1$. Let us illustrate by some examples.

Illustration. Prove that:

$$1 + 3 + 5 + 7 + \cdots + (2n - 1) = n^2.$$

Solution. First we show that it is true for particular values of n.

$$n = 1, \quad 1 = 1^2 \qquad \text{or} \quad 1 = 1.$$
$$n = 2, \quad 1 + 3 = 2^2 \qquad \text{or} \quad 4 = 4.$$
$$n = 3, \quad 1 + 3 + 5 = 3^2 \qquad \text{or} \quad 9 = 9.$$

The second part assumes it to be true for $n = k$, so that we have:

$$1 + 3 + 5 + \cdots + (2k - 1) = k^2.$$

Now add the next odd integer, $2k + 1$, to both sides.

$$1 + 3 + 5 + \cdots + (2k - 1) + (2k + 1) = k^2 + (2k + 1) = (k + 1)^2.$$

Thus we see that it is true for $n = k + 1$ if it is true for $n = k$. But we know that it is true for $n = 1$, $n = 2$, and $n = 3$ by our verification. Consequently, it is true for $3 + 1 = 4$, $4 + 1 = 5$, etc.

Binomial Theorem. *The binomial formula for $(a + b)^n$ holds for all positive integral values of n.*

Proof. We have already shown that it holds for $n = 1,2,3,4$, and 5. Next, we assume it holds for $n = k$ so that:

$$(a + b)^k = a^k + ka^{k-1}b + \frac{k(k - 1)}{1 \cdot 2} a^{k-2}b^2 + \cdots$$
$$+ \frac{k(k - 1) \cdots (k - r + 1)}{r!} a^{k-r}b^r + \cdots + b^k.$$

Now multiply both members by $(a + b)$; for the left member we obtain $(a + b)^{k+1}$; for the right member:

$$a^{k+1}+ka^kb \quad + \cdots + \frac{k(k-1) \cdots (k-r+1)}{r!} a^{k-r+1}b^r \quad + \cdots$$

$$+a^kb \quad + \cdots + \frac{k(k-1) \cdots (k-r+2)}{(r - 1)!} a^{k-r+1}b^r \quad + \cdots +b^{k+1}$$

$$\overline{a^{k+1}+(k+1)a^kb+ \cdots + \frac{(k+1)k(k-1) \cdots (k-r+2)}{r!} a^{k-r+1}b^r+ \cdots +b^{k+1}}$$

The first line is the multiplication by a, the second by b, and the third line is the sum of the two. The addition to obtain the coefficient of $a^{k-r+1}b^r$ may be troublesome, so let us show it in detail. We need to simplify:

$$\frac{k(k - 1) \cdots (k - r + 1)}{r!} + \frac{k(k - 1) \cdots (k - r + 2)}{(r - 1)!}.$$

First factor out the common factor:

$$\frac{k(k - 1) \cdots (k - r + 2)}{(r - 1)!}.$$

This leaves:

$$\frac{k - r + 1}{r} \quad \text{and} \quad 1$$

so that we have:

$$\frac{k - r + 1}{r} + 1 = \frac{k - r + 1 + r}{r} = \frac{k + 1}{r}.$$

Combining this with the common factor yields:

$$\frac{k(k - 1) \cdots (k - r + 2)}{(r - 1)!} \cdot \frac{k + 1}{r} = \frac{(k + 1)k(k - 1) \cdots (k - r + 2)}{r!}.$$

We see that the resulting expression is exactly the same as the binomial formula with k replaced by $k + 1$. Thus since the formula is true for $k = 1, 2, 3, 4$, and 5 it is true for $5 + 1 = 6$, etc.

45. The Binomial Formula for n Not a Positive Integer. If we apply the binomial formula to a binomial and have n equal to a number which is not a positive integer, we obtain a never-ending series of terms known as the **binomial series**. This is the student's first introduction to a large branch of mathematics known, in general, as *series*. They are studied quite extensively in advanced mathematics and find application in many problems. In this book we shall simply illustrate the binomial series.

Illustration. Find the first three terms of the expansion of $(x^3 - \frac{1}{2}y)^{\frac{1}{3}}$.

Solution. Applying the binomial formula:

$$\left(x^3 - \frac{1}{2}y\right)^{\frac{1}{3}} = (x^3)^{\frac{1}{3}} + \frac{1}{3}(x^3)^{-\frac{2}{3}}\left(-\frac{1}{2}y\right) + \frac{(\frac{1}{3})(-\frac{2}{3})}{1\cdot 2}(x^3)^{-\frac{5}{3}}\left(-\frac{1}{2}y\right)^2 + \cdots$$

$$= x - \tfrac{1}{6}x^{-2}y - \tfrac{1}{36}x^{-5}y^2 + \cdots$$

$$= x - \frac{y}{6x^2} - \frac{y^2}{36x^5} + \cdots.$$

Illustration. Find the value of $\sqrt{1.03}$ correct to four decimal places.

Solution. Write:

$$(1.03)^{\frac{1}{2}} = (1 + .03)^{\frac{1}{2}}$$

$$= 1^{\frac{1}{2}} + \frac{1}{2}(1)^{-\frac{1}{2}}(.03) + \frac{(\frac{1}{2})(-\frac{1}{2})}{1\cdot 2}(1)^{-\frac{3}{2}}(.03)^2$$

$$+ \frac{(\frac{1}{2})(-\frac{1}{2})(-\frac{3}{2})}{1\cdot 2\cdot 3}(1)^{-\frac{5}{2}}(.03)^3 + \cdots$$

$$= 1 + .015 - \tfrac{1}{8}(.0009) + \tfrac{1}{16}(.000027) - \cdots$$

$$= 1 + .015 - .0001125 + .0000016875$$

$$= 1.01489 = 1.0149.$$

Illustration. Find the first four terms of $\dfrac{1}{1 + x}$.

Solution. Write:

$$(1 + x)^{-1} = 1^{-1} + (-1)(1)^{-2}x + \frac{(-1)(-2)}{1\cdot 2}(1)^{-3}x^2$$

$$+ \frac{(-1)(-2)(-3)}{1\cdot 2\cdot 3}(1)^{-4}x^3 + \cdots$$

$$= 1 - x + x^2 - x^3 + \cdots.$$

46. Inequalities. An inequality is simply a statement that one expression *is greater than or less than another*. We have seen the symbol

$a > b$, which reads "a is greater than b," and $a < b$, which reads "a is less than b." There are many ways in which to make these statements. For example, there are three ways of expressing the statement "a is greater than b":

(i) $a > b$ or $b < a$.
(ii) $a - b > 0$; $a - b$ is a positive number.
(iii) $a - b = n$; n is a positive number.

If an expression *is either greater than or equal to*, we use the symbol $\geq$, and similarly, $\leq$ states *is less than or equal to*. Two inequalities are *alike in sense*, or of the *same sense*, if their symbols for inequality point in the same direction. Similarly, they are *unlike*, or *opposite in sense*, if the symbols point in opposite directions.

In discussing inequalities of algebraic expressions we see that we can have two classes of them.

1. If the sense of the inequality is the same for **all** values of the symbols for which its members are defined, the inequality is called an **absolute** or **unconditional inequality**.

Illustrations.

$$x^2 + y^2 > 0, \quad x \neq 0 \quad \text{or} \quad y \neq 0.$$
$$\pi < 4.$$

2. If the sense of the inequality holds only for **certain** values of the symbols involved, the inequality is called a **conditional inequality**.

Illustrations.

$$x + 3 < 7, \quad \text{true only for values of } x \text{ less than 4};$$
$$x^2 + 6 < 5x, \text{ true only for } x \text{ between 2 and 3}.$$

The inequality symbols are frequently used to denote the values of a variable between given limits. Thus, $1 \leq x < 4$, states "values of x from 1, including 1, to 4 but not including 4"; i.e., x may assume the values 1 and from 1 to 4 but no others. This is also called "defining the **range of values**."

Illustration.

$$x^2 + 6 < 5x \quad \text{for} \quad 2 < x < 3.$$

Properties.

I. The sense of an inequality is *not* changed if both members are increased or diminished by the *same* number. If $a > b$, then:

$$a + x > b + x \quad \text{and} \quad a - x > b - x.$$

II. If $a > b$ and $x > 0$, then:

$$ax > bx \quad \text{and} \quad \frac{a}{x} > \frac{b}{x}.$$

III. If $a > b$ and $x < 0$, then:

$$ax < bx \quad \text{and} \quad \frac{a}{x} < \frac{b}{x}.$$

IV. If a, b, and n are positive numbers and $a > b$, then:

$$a^n > b^n \quad \text{and} \quad \sqrt[n]{a} > \sqrt[n]{b}.$$

V. If $x > 0$, $a > b$, and a and b have like signs, then:

$$\frac{x}{a} < \frac{x}{b}.$$

We can illustrate these properties by using numbers.

Illustrations.

(1) Since $4 > 3$, we have $4 + 2 > 3 + 2$ as $6 > 5$.
(2) Since $4 > 3$, we have $4(2) > 3(2)$ as $8 > 6$.
(3) Since $4 > 3$, we have $4(-2) < 3(-2)$ as $-8 < -6$.
(4) Since $16 > 9$, we have $\sqrt{16} > \sqrt{9}$ as $4 > 3$.
(5) Since $4 > 3$, we have $\frac{2}{4} < \frac{2}{3}$ as $\frac{1}{2} < \frac{2}{3}$.

The solutions of inequalities are obtained in a manner very similar to that of obtaining solutions to equations. The main difference is that we are now finding a *range* of values of the unknown such that the inequality is satisfied. Furthermore, we must pay strict attention to the properties so that in performing operations we do not change the sense of the inequality without knowing it.

Illustration. Solve the inequality:

$$7x - 3 > 2x + 7.$$

Solution. Using the properties we write in turn:

$$7x - 3 > 2x + 7.$$
$$7x - 2x > 7 + 3.$$
$$5x > 10.$$
$$x > 2.$$

The solution, then, is all values of x greater than 2. Graphically the range is pictured in Fig. 23.

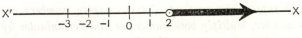

FIG. 23

47. Complex Numbers. In Section 12 we defined an imaginary number. This number came into existence when we attempted to obtain the square root of a negative number. We then defined the *imaginary unit* by:

$$i^2 = -1$$

and displayed the properties of this number. By the use of it we were able to find the square root of any negative number by expressing it as a product of a real number and the imaginary unit; thus:

$$\sqrt{-16} = i\sqrt{16} = 4i.$$

Any number of the form *bi* is called a **pure imaginary number.** If we combine it with a real number by an algebraic sign, *a + bi*, we have defined a **complex number.** The number *a* is the **real part,** and the number *bi* is the **imaginary part.** It is easily seen that the real-number system and the pure-imaginary system are special cases of the complex-number system.

Complex numbers that differ only in the signs of their imaginary part are called **conjugate complex numbers.** Thus, *a − bi* is the conjugate of *a + bi*, and in turn *a + bi* is the conjugate of *a − bi*.

Two complex numbers are said to be equal if and only if their real parts are equal and their imaginary parts are equal.

Thus:

$$a + bi = x + yi \quad \text{if and only if} \quad a = x \quad \text{and} \quad b = y;$$
$$a + bi = 0 \quad\quad \text{if and only if} \quad a = b = 0.$$

Operations.

I. *To add or subtract two complex numbers, add or subtract the real and imaginary parts separately:*

$$(a + bi) + (c + di) = (a + c) + (b + d)i.$$
$$(a + bi) - (c + di) = (a - c) + (b - d)i.$$

II. *To find the product of two complex numbers, multiply them together as two binomials and substitute* −1 *for* i^2:

$$(a + bi)(c + di) = ac + adi + bci + bdi^2$$
$$= (ac - bd) + (ad + bc)i.$$

III. *To express the quotient of two complex numbers as a single complex number, multiply both numerator and denominator by the conjugate of the denominator:*

$$\frac{a + bi}{c + di} = \frac{(a + bi)(c - di)}{(c + di)(c - di)} = \frac{ac + bd}{c^2 + d^2} + \frac{bc - ad}{c^2 + d^2}\, i.$$

As in the real-number system, division by zero is *not* permitted in the complex-number system. We note that the product of two conjugate complex numbers is a real number.

$$(a + bi)(a - bi) = a^2 - (bi)^2 = a^2 - b^2 i^2 = a^2 + b^2.$$

Division of complex numbers can be thought of as an operation similar to that of rationalizing the denominator.

Illustrations.

(1) $(4 - 6i) + (3 + 2i) = 7 - 4i.$

(2) $(3 + 2i) - (7 + 4i) = -4 - 2i.$

(3) $(-4 + 7i)(3 - 2i) = [-12 - (-14)] + (8 + 21)i = 2 + 29i.$

(4) $\dfrac{6 - i}{1 + 3i} = \dfrac{(6 - i)(1 - 3i)}{(1 + 3i)(1 - 3i)} = \dfrac{3}{10} - \dfrac{19}{10}\, i.$

48. Trigonometric Form of a Complex Number. The graphical representation of a complex number can be accomplished by locating the real numbers a and b on a rectangular coordinate system. Let us choose the X-axis as the real axis and the Y-axis as the imaginary axis. The point, $P(a + bi)$, is then located as in Fig. 24.

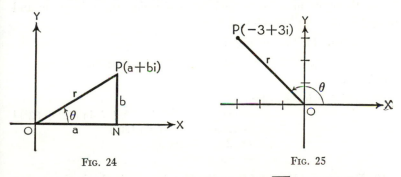

FIG. 24   FIG. 25

If we join the point P to the origin and denote $\overline{OP}$ by r and the angle this line segment makes with the positive X-axis by θ, we can then verify the following relations:

$$a = r \cos \theta, \qquad b = r \sin \theta,$$
$$r = \sqrt{a^2 + b^2}, \quad \tan \theta = \frac{b}{a}.$$

Thus we have:

$$a + bi = r(\cos \theta + i \sin \theta).$$

This is the **trigonometric form** of the complex number; r is called the **modulus** and is always positive; the angle θ is called the **argument** or **amplitude**.

Illustration. Change $-3 + 3i$ to its trigonometric form.

Solution. See Fig. 25, p. 75.

$$r = \sqrt{(-3)^2 + (3)^2} = 3\sqrt{2}.$$

$$\tan \theta = \frac{3}{-3} = -1.$$

Since a is negative and b is positive, the angle is in the second quadrant;

$$\theta = 135°.$$

$$\therefore \ -3 + 3i = 3\sqrt{2}(\cos 135° + i \sin 135°).$$

Theorem 1.

$$r_1(\cos \theta_1 + i \sin \theta_1) \cdot r_2 (\cos \theta_2 + i \sin \theta_2)$$
$$= r_1 r_2 [\cos (\theta_1 + \theta_2) + i \sin (\theta_1 + \theta_2)].$$

Theorem 2.

$$\frac{r_1(\cos \theta_1 + i \sin \theta_1)}{r_2(\cos \theta_2 + i \sin \theta_2)} = \frac{r_1}{r_2}[\cos (\theta_1 - \theta_2) + i \sin (\theta_1 - \theta_2)].$$

These theorems may be proved by performing the arithmetic and using the addition formulas of Section 34.

Illustration. Find the product:

$$3(\cos 38° + i \sin 38°) \cdot 4(\cos 82° + i \sin 82°).$$

Solution.

$$3(\cos 38° + i \sin 38°) \cdot 4(\cos 82° + i \sin 82°)$$
$$= (3)(4)[\cos (38° + 82°) + i \sin (38° + 82°)]$$
$$= 12(\cos 120° + i \sin 120°)$$
$$= 12(-\tfrac{1}{2} + \tfrac{1}{2}i\sqrt{3})$$
$$= -6 + 6i\sqrt{3}.$$

49. Powers and Roots of Complex Numbers. To obtain the powers and roots of complex numbers we employ a famous theorem known as **De Moivre's Theorem.**

If n is any positive integer, then:

$$(\cos \theta + i \sin \theta)^n = \cos n\theta + i \sin n\theta.$$

The theorem is proved by mathematical induction and using the property of multiplication of the last section. It can also be shown that the theorem is true when n is a fraction or a negative number. We may now use this theorem to find the powers of a complex number.

Illustration. Find $(-3 + 3i)^4$.

Solution. By the aforementioned properties:

$$\begin{aligned}
(-3 + 3i)^4 &= [3\sqrt{2}(\cos 135° + i \sin 135°)]^4 \\
&= (3\sqrt{2})^4(\cos 540° + i \sin 540°) \\
&= 324(\cos 180° + i \sin 180°) \\
&= 324(-1 + i \cdot 0) \\
&= -324.
\end{aligned}$$

To find the roots of a complex number we first note that:

$$r(\cos \theta + i \sin \theta) = r[\cos (\theta + k360°) + i \sin (\theta + k360°)]$$

because of the periodicity of the trigonometric functions. Now applying De Moivre's Theorem we have:

$$[r(\cos \theta + i \sin \theta)]^{\frac{1}{n}} = \sqrt[n]{r} \cos \frac{\theta + k360°}{n} + i \sin \frac{\theta + k360°}{n}.$$

Illustration. Find $\sqrt[3]{9 + 3i\sqrt{3}}$.

Solution.

$$r = \sqrt[2]{9^2 + (3\sqrt{3})^2} = \sqrt[2]{81 + 27} = 6\sqrt{3}.$$

$$\tan \theta = \frac{3\sqrt{3}}{9} = \frac{1}{3}\sqrt{3}; \; \theta = 30°.$$

Thus:

$$9 + 3i\sqrt{3} = 6\sqrt{3}(\cos 30° + i \sin 30°)$$

and

$$\begin{aligned}
(9 + 3i\sqrt{3})^{\frac{1}{3}} &= (6\sqrt{3})^{\frac{1}{3}} \left[\cos \frac{30° + k360°}{3} + i \sin \frac{30° + k360°}{3} \right] \\
&= (108)^{\frac{1}{6}}[\cos (10° + k120°) + i \sin (10° + k120°)].
\end{aligned}$$

The three cube roots are:

$$\begin{aligned}
k = 0: \; &(108)^{\frac{1}{6}}(\cos 10° + i \sin 10°); \\
k = 1: \; &(108)^{\frac{1}{6}}(\cos 130° + i \sin 130°); \\
k = 2: \; &(108)^{\frac{1}{6}}(\cos 250° + i \sin 250°).
\end{aligned}$$

50. Exercise IV.

1. A particle is constrained to move on a circle by a force which varies directly as the square of the velocity, V, and inversely as the radius, r. If the force is 108 when V is $4\frac{1}{2}$ and $r = 3$, what is the force when V is 5 and $r = 4$?

2. Find the first four terms in the expansions of:

   (a) $(3x - 2y)^{\frac{1}{3}}$.

   (b) $(x^{\frac{1}{2}} - 2y^{\frac{1}{3}})^{-2}$.

   (c) $(ax + by)^5$.

3. Find the 7th term in the expansion of $(2x^{\frac{2}{3}} - 3y^{\frac{1}{2}})^{-\frac{1}{2}}$.

4. Find the range of values for which:

   (a) $9x + 7 > 2x + 21$.

   (b) $4x - \sqrt{32} > x - 4\sqrt{2}$.

5. Simplify the following complex numbers:

   (a) $3 - 6i + 7 - 8 + i + 7i - 1 - i$.

   (b) $(-3 - 2i)(-5 - 3i)$.

   (c) $(9 + 9i) \div (3 - 3i)$.

6. Change the following complex numbers to trigonometric form:

   (a) $\sqrt{2} + \sqrt{2}i$.

   (b) $12 - 5i$.

7. Simplify:

   (a) $5(\cos 12° + i \sin 12°) \cdot 4(\cos 78° + i \sin 78°)$.

   (b) $12(\cos 138° + i \sin 138°) \div 6(\cos 93° + i \sin 93°)$.

8. Find:

   (a) $(1 - i\sqrt{3})^5$.

   (b) $(2 - 2i\sqrt{3})^{\frac{1}{4}}$.

5

THE LINEAR FUNCTION

51. Introduction. In this chapter we shall discuss the solutions of linear equations and the representation of a linear function. We are already familiar with the definition of a linear function, as one in which the highest degree of the independent variable is one, and if we let $y = f(x)$ we say that the most general expression for a linear function is:

$$Ax + By + C.$$

There are many other forms and some interesting applications to which we shall now turn.

If we set a linear algebraic expression in one unknown equal to zero, we have the simple linear equation:

$$ax + b = 0.$$

If we set the left member equal to y, we have the simple linear function:

$$y = ax + b.$$

By giving values to x and finding the corresponding values of y we obtain a set of coordinates which, when plotted on coordinate axes, yield a straight line. Thus the representation of a simple linear function is a *straight line*.

The **zero** of the linear function is the value of the independent variable for which the corresponding value of the function is zero. Or in the function:

$$y = ax + b$$

the zero of the function is the value of x for which y equals zero. Thus in the illustration on p. 80 the zero is at $x = 2$ and this is the solution of the equation $2x - 4 = 0$.

Illustration. Plot $y = 2x - 4$.

Solution. Give values to x and obtain the coordinates.

x	y
-1	-6
0	-4
1	-2
2	0
3	2
4	4
5	6

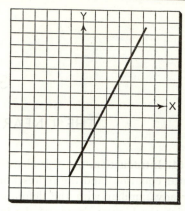

FIG. 26

52. Division of a Line Segment. A line segment is defined by two points on a line, say $P_1(x_1,y_1)$ and $P_2(x_2,y_2)$. Let it be desired to find a point $P(x,y)$ lying on the line and such that the portion P_1P is a given fraction of the P_1P_2 segment. By the use of ratios we write:

$$P_1P = k(P_1P_2).$$

Consider the configuration of Fig. 27. By similar triangles we have:

$$\frac{N_1N}{N_1N_2} = \frac{P_1P}{P_1P_2} = k,$$

so that:

$$N_1N = kN_1N_2.$$

Now:

$$N_1N = x - x_1 \text{ and } N_1N_2 = x_2 - x_1.$$
$$\therefore \quad x - x_1 = k(x_2 - x_1)$$
$$\text{or} \quad x = x_1 + k(x_2 - x_1).$$

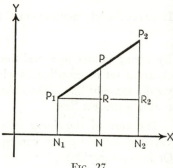

FIG. 27

Similarly, for y we have:

$$y = y_1 + k(y_2 - y_1).$$

Illustration. Find the first point of trisection of the line from $(7,2)$ to $(1,-7)$.

Solution.

$$k = \tfrac{1}{3}.$$

$$x = x_1 + k(x_2 - x_1)$$
$$= 7 + \tfrac{1}{3}(1 - 7) = 7 - 2 = 5.$$
$$y = y_1 + k(y_2 - y_1)$$
$$= 2 + \tfrac{1}{3}(-7 - 2) = 2 - 3 = -1.$$
$$\therefore \quad P(x,y) = P(5,-1).$$

A special case is the *midpoint*, for which $k = \tfrac{1}{2}$. Thus the coordinates (x,y) of the midpoint between (x_1,y_1) and (x_2,y_2) are:

$$x = \tfrac{1}{2}(x_1 + x_2), \quad y = \tfrac{1}{2}(y_1 + y_2).$$

Illustration.   Find the coordinate of the midpoint between $(3,-2)$ and $(-4,2)$.

Solution.

$$x = \tfrac{1}{2}(x_1 + x_2) = \tfrac{1}{2}(3 - 4) = -\tfrac{1}{2}.$$
$$y = \tfrac{1}{2}(y_1 + y_2) = \tfrac{1}{2}(-2 + 2) = 0.$$

Draw the line segment and check.

53. Slope of a Line.   The acute angle between a straight line and the positive X-axis is called the *angle of inclination*. The *tangent of the angle of inclination* is called the **slope** of the line. If we denote the slope by m, we may write:

$$m = \tan \alpha.$$

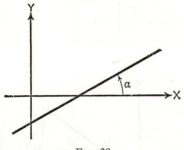

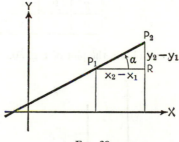

Fig. 28	Fig. 29

The slope is *positive* if α is a *positive* acute angle; otherwise it is negative.

If we choose two points $P_1(x_1,y_1)$ and $P_2(x_2,y_2)$ on a line we see that:

$$m = \tan \alpha = \frac{RP_2}{P_1R}$$

or

$$m = \frac{y_2 - y_1}{x_2 - x_1}.$$

Thus the slope is a ratio of the vertical distance to the horizontal distance. A line parallel to the X-axis has $y_2 = y_1$, so that $y_2 - y_1 = 0$ and the slope is zero. A line parallel to the Y-axis has $x_2 = x_1$ and $x_2 - x_1 = 0$; this would give rise to a division by zero, which is not defined. Thus the slope of a line parallel to the Y-axis is undefined, although some authors call this slope infinite.

54. Forms Defining a Straight Line. There are many different ways in which we can write the equation of a straight line. We shall now list some of the more important ones.

I. *Parallel to a coordinate axis.* If a straight line is parallel to the Y-axis, its equation is:

(1a) $$x = k.$$

If a straight line is parallel to the X-axis, its equation is:

(1b) $$y = k.$$

These forms are clear if we consider the equations to read that $x = k$ (or $y = k$) for all values of y (or x).

Illustration. Plot the line $x = 3$.

Solution. We let $x = 3$ for all values of y, thus:

y	0	1	2	3	4	etc.
x	3	3	3	3	3	

and obtain the plot of Fig. 30.

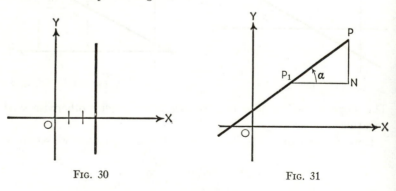

FIG. 30   FIG. 31

II. *Point-slope form.* The equation of a straight line passing through a given point $P_1(x_1, y_1)$ having a given slope, m, is:

(2) $$y - y_1 = m(x - x_1).$$

Proof.   Construct Fig. 31, where $P(x,y)$ is any point on the line through $P_1(x_1,y_1)$. We note that:

$$\tan \alpha = \frac{NP}{P_1N} = \frac{y - y_1}{x - x_1} = m.$$

or $$y - y_1 = m(x - x_1).$$

Illustration.   Find the equation of the line with a slope of -2 and passing through $(3,1)$.

Solution.

$$y - 1 = -2(x - 3).$$
$$y = -2x + 7.$$

III. *Two-point form.*   The equation of a straight line through two given points $P_1(x_1,y_1)$ and $P_2(x_2,y_2)$ is:

(3) $$y - y_1 = \frac{y_2 - y_1}{x_2 - x_1}(x - x_1).$$

This form is obtained by simply substituting for the slope in equation (2) since by Section 53:

$$m = \frac{y_2 - y_1}{x_2 - x_1}.$$

If $x_2 = x_1$ the slope is undefined as $x_2 - x_1$ becomes zero. However, formula (3) may be written in the form:

(3a) $$(y - y_1)(x_2 - x_1) = (y_2 - y_1)(x - x_1).$$

Then, if $x_2 = x_1$, we have:

$$0 = (y_2 - y_1)(x - x_1) \quad \text{or} \quad x = x_1,$$

which is a line parallel to the Y-axis.

Illustration.   Find the equation of a line through the points $(-3,2)$ and $(2,-5)$.

Solution.   We have:

$$y_2 - y_1 = -5 - 2 = -7,$$
$$x_2 - x_1 = 2 - (-3) = 5.$$

Substituting into formula (3) we obtain:

$$y - 2 = -\tfrac{7}{5}(x + 3)$$
or $$5y - 10 = -7x - 21.$$
$$5y + 7x + 11 = 0.$$

Check.

$$5(2) + 7(-3) + 11 = 0; \quad 5(-5) + 7(2) + 11 = 0.$$

Note that it is immaterial which point we pick as P_1 or P_2 as long as we are consistent. Reverse the order of the points in the illustration and solve the problem.

IV. *Slope-intercept form.* The equation of a straight line whose slope is m and whose y-intercept (see Section 19) is b is:

$$(4) \qquad\qquad y = mx + b.$$

The point where the line crosses the Y-axis is $P_1(0,b)$. Substituting this point into formula (2) we obtain:

$$y - b = m(x - 0)$$
or
$$y = mx + b.$$

This is the usual linear function form; see Section 52.

Illustration. Find the equation of a line with slope equal to -2 and crossing the Y-axis at $y = 3$.

Solution. We have $m = -2$ and $b = 3$; thus:

$$y = -2x + 3$$
or
$$y + 2x = 3.$$

Check. $P_1(0,3)$: $\qquad 3 + 2(0) = 3.$

V. *The general form.* The general form of the equation of a straight line is given by:

$$(5) \qquad\qquad Ax + By + C = 0.$$

This is sometimes expressed by the following theorem.

Theorem. *The locus of every equation of the first degree is a straight line,* and conversely, *the equation of a straight line is always of the first degree.*[*]

We note some properties:

(i) If $A = 0$, the line is parallel to the X-axis.
(ii) If $B = 0$, the line is parallel to the Y-axis.
(iii) If we solve for y, we obtain the slope-intercept form and can read off the slope and y-intercept.

$$y = -\frac{A}{B}x - \frac{C}{B}.$$

---

[*] If the locus of a higher-degree equation is a straight line, the equation can be factored into linear factors (at least one), each of which (the one) can be considered to be the equation of "the" line.

Thus:
$$m = -\frac{A}{B} \text{ and } b = -\frac{C}{B}.$$

VI. *Intercept form.* The equation of a straight line with intercepts a and b is:

(6)
$$\frac{x}{a} + \frac{y}{b} = 1.$$

Let us find the equation of a line through the points $P_1(a,0)$ and $P_2(0,b)$. By formula (3):

$$y - 0 = \frac{b}{-a}(x - a).$$

Simplify to get:

$$ay = -bx + ab$$
or
$$ay + bx = ab.$$

Divide by ab to obtain formula (6), which is called the *intercept form.* It is clear that this form fails if the line passes through the origin, $(0,0)$. If we know the intercepts we have an easy method of drawing the line by simply joining the points $(a,0)$ and $(0,b)$.

Illustration. Find the intercepts and draw the line for the equation:

$$3x + 2y = 6.$$

Solution. To obtain the intercepts let $x = 0$ and solve for y; then let $y = 0$ and solve for x.

$$x = 0, \quad 2y = 6 \quad \text{or} \quad y = 3 = b.$$
$$y = 0, \quad 3x = 6 \quad \text{or} \quad x = 2 = a.$$

Locate the points $(2,0)$ and $(0,3)$ and join them with a straight line.

55. Parallel and Perpendicular Lines. In Section 53 we defined the slope of a line to be:

$$m = \tan \alpha.$$

We recall from elementary geometry that if two lines are parallel they make the same angle with a transversal such as the X-axis. Thus in Fig. 32, p. 86, $\alpha_1 = \alpha_2$ and we see that the **slopes of two parallel lines are equal.**

Theorem. *If two lines are perpendicular, their slopes are negative reciprocals of each other, and conversely.* This can be expressed by:

$$m_1 m_2 = -1.$$

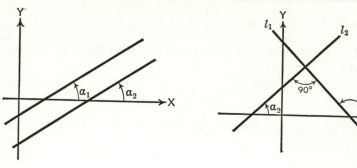

FIG. 32   FIG. 33

Proof. Consider Fig. 33. From elementary geometry we have:

$$\alpha_1 = 90° + \alpha_2.$$

Thus:

$$m_1 = \tan \alpha_1 = \tan (90° + \alpha_2)$$

$$= -\cot \alpha_2 = -\frac{1}{\tan \alpha_2}$$

$$= -\frac{1}{m_2}.$$

In the general form these two properties are expressed by:

$$\begin{cases} Ax + By + C = 0 \\ Ax + By + D = 0 \end{cases}$$

are parallel, and:

$$\begin{cases} Ax + By + C = 0 \\ Bx - Ay + D = 0 \end{cases}$$

are perpendicular.

Illustration. Find the equation of a line through the point $(2,-3)$ and perpendicular to the line $2x - y - 3 = 0$.

Solution. The coefficients of x and y of the desired equation are obtained by interchanging the coefficients of x and y of the given equation and changing the sign of one of them; thus:

$$x + 2y = c.$$

To determine c substitute the coordinates of the given point, $(2,-3)$:

$$2 + 2(-3) = c = 2 - 6 = -4.$$

The desired equation is:

$$x + 2y + 4 = 0.$$

56. Angle between Two Lines. In discussing the angle between two lines we shall understand the *acute angle*, θ, through which the line l_1 must be rotated to coincide with the line l_2. The angle is *positive* if measured counterclockwise and *negative* if measured clockwise. Consider Fig. 34 in which we picture the given lines l_1 and l_2 making angles α_1 and α_2 with the X-axis. By elementary geometry we have:

$$\alpha_2 = \alpha_1 + \theta$$

or
$$\theta = \alpha_2 - \alpha_1.$$

Thus:
$$\tan \theta = \tan (\alpha_2 - \alpha_1) = \frac{\tan \alpha_2 - \tan \alpha_1}{1 + \tan \alpha_2 \tan \alpha_1}$$

or
$$\boldsymbol{\tan \theta = \frac{m_2 - m_1}{1 + m_1 m_2}}.$$

This is the formula which will determine the angle from l_1 to l_2.

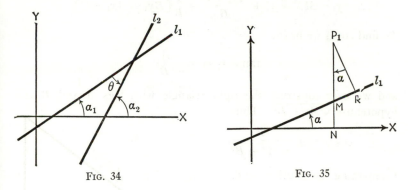

FIG. 34   FIG. 35

Illustration. Find the angle from the line $y - 3x - 4 = 0$ to the line $y + 2x + 1 = 0$.

Solution. Find the slopes of each line.

$$m_1 = 3 \quad \text{and} \quad m_2 = -2.$$

Thus:
$$\tan \theta = \frac{-2 - 3}{1 + (-2)(3)} = \frac{-5}{-5} = 1.$$

$$\therefore \quad \theta = 45°.$$

57. Distance from a Point to a Line. Let us consider an interesting problem in geometry. Given a line, l_1:

$$Ax + By + C = 0$$

and a point $P_1(x_1,y_1)$ not on the line. It is desired to find the perpendicular distance from P_1 to the line. Draw Fig. 35 showing the line l_1 with slope tan α and the point P_1. Draw the $\perp P_1MN$ from P_1 to the X-axis. By elementary geometry $\angle NP_1R = \alpha$. The desired distance:

$$d = P_1R = P_1M \cos \alpha = (P_1N - MN) \cos \alpha$$
$$= (y_1 - MN) \cos \alpha.$$

To find MN consider the point $M(x_1,y)$ whose coordinates must satisfy the equation of the line; thus:

$$Ax_1 + By + C = 0$$

or

$$y = -\frac{Ax_1 + C}{B}, \quad \text{if } B \neq 0.$$

$$\therefore \quad y_1 - MN = y_1 + \frac{Ax_1 + C}{B} = \frac{1}{B}[By_1 + Ax_1 + C].$$

To find $\cos \alpha$ we have:

$$\tan \alpha = m = -\frac{A}{B},$$

and we may construct the right triangle with legs A and B and hypotenuse $\sqrt{A^2 + B^2}$. Thus:

$$\cos \alpha = \pm \frac{B}{\sqrt{A^2 + B^2}}.$$

Completing the solution for d:

$$(y_1 - MN) \cos \alpha =$$
$$\frac{1}{B}[By_1 + Ax_1 + C]\left[\pm \frac{B}{\sqrt{A^2 + B^2}}\right]$$

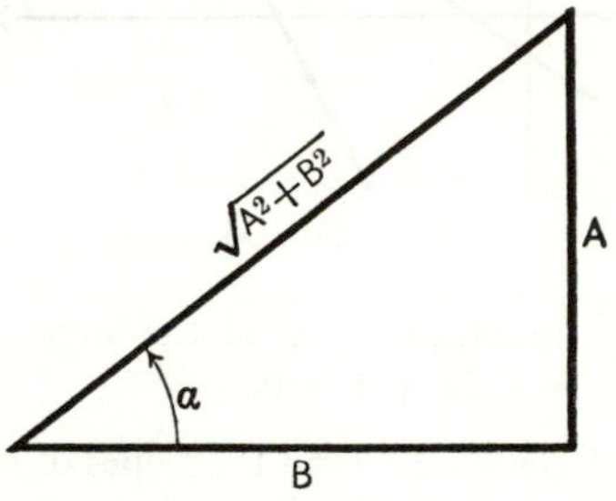

FIG. 36

or

$$d = \pm \frac{Ax_1 + By_1 + C}{\sqrt{A^2 + B^2}}.$$

The sign of $\pm$ is usually chosen to make d positive.

Illustration. Find the distance from $P(2,5)$ to the line $3x + 4y - 6 = 0$.
Solution. $\sqrt{A^2 + B^2} = \sqrt{9 + 16} = 5$.

$$Ax_1 + By_1 + C = 3(2) + 4(5) - 6 = 20.$$
$$d = \frac{20}{5} = 4.$$

58. Rate of Change of a Linear Function. In Section 16 we pointed out that the rate of change of y in one of the illustrations was constant and stated that this was characteristic of a linear function. The precise statement is given in the form of a theorem.

Theorem. *The rate of change of a linear function is constant and equal to the slope of its straight line.*

Proof. By definition we have:

$$\Delta x = x_2 - x_1, \quad \Delta y = y_2 - y_1,$$

and $\quad \dfrac{\Delta y}{\Delta x} =$ rate of change of y with respect to x.

However, for two points on a given line we have:

$$m = \frac{y_2 - y_1}{x_2 - x_1} = \frac{\Delta y}{\Delta x}.$$

Thus the rate of change of a linear function is equal to the slope of a straight line, which is constant.

We should also notice that a positive rate means the function is *increasing* as the independent variable *increases;* negative rate, *decreasing.* Draw a straight line with positive slope and notice how y increases as x increases. Draw a straight line with negative slope and investigate y as x increases.

59. System of Linear Equations. The linear equation:

$$Ax + By + C = 0$$

has an unlimited number of solutions. In fact, every point on the straight line which represents this equation satisfies the equation. Let us now consider two linear equations in the same two unknowns.

$$\begin{cases} A_1x + B_1y + C_1 = 0, \\ A_2x + B_2y + C_2 = 0. \end{cases}$$

These two equations are said to form a **system of simultaneous linear equations.** A **solution** of this system is a pair of corresponding values of x and y which satisfy both of these equations at the same time. There are three things which can happen in the study of this simple system. Since each of the equations has for its graph a straight line, we may discuss the possibilities in terms of the straight lines.

(i) If the straight lines have only one point in common, the coordinates of this point constitute a solution of the system, and the equations are said to be **consistent.**

(ii) If the two lines are parallel, they have no point in common, there is no solution, and the equations are said to be **inconsistent**.

(iii) If the two lines are coincident, all points are common, there is an unlimited number of solutions, and the equations are said to be **dependent**.

The three conditions can easily be checked by considering the slopes of the lines.

If the slopes are **equal,** *the lines are either* **parallel** *or* **coincident.** Another way to say the same thing is:

If the slopes are **not equal,** *the system has one and only one solution.*

Illustrations. Determine the nature of the following systems of linear equations:

(1) $\begin{cases} 3x + 2y - 6 = 0, \\ x - y + 2 = 0. \end{cases}$

Solution. $m_1 = -\frac{3}{2}, m_2 = 1.$ ∴ Consistent.

(2) $\begin{cases} x + 5y = 4, \\ 3x + 15y = 12. \end{cases}$

Solution. $m_1 = -\frac{1}{5}, m_2 = -\frac{3}{15} = -\frac{1}{5}.$ ∴ Inconsistent or dependent.

(3) $\begin{cases} x + y + 3 = 0, \\ x + y - 6 = 0. \end{cases}$

Solution. $m_1 = -1, m_2 = -1.$ ∴ Inconsistent or dependent.

60. Solutions of Systems of Two Linear Equations. There are a number of methods of finding the solutions of systems of linear equations. In this section we shall be primarily concerned with two algebraic methods. The graphical method will yield approximate solutions but should be used only to give a rough check. Before starting any solution check the slopes to determine whether or not to expect a solution.

The graphical method consists simply of plotting the two straight lines and reading off the point of intersection.

Illustration. Solve:

$$\begin{cases} y + 2x = 7, \\ 2y + 2x = 10. \end{cases}$$

Solution. Obtain the intercepts:

$$a_1 = \tfrac{7}{2}, b_1 = 7.$$
$$a_2 = 5, b_2 = 5.$$

Draw the lines, Fig. 37.

Point of intersection: $x = 2$, $y = 3$.

Check. $\begin{cases} 3 + 2(2) = 7 = 7, \\ 2(3) + 2(2) = 10 = 10. \end{cases}$

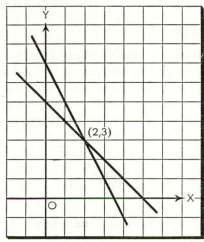

FIG. 37

We will now discuss two methods of **elimination**: by **addition or subtraction** and by **substitution**.

Illustration 1. Solve:

$$\begin{cases} 2x - 4y + 4 = 0, & (1) \\ 3x + 2y + 8 = 0. & (2) \end{cases}$$

Solution.

Write (1): $\qquad\qquad\qquad 2x - 4y + 4 = 0.$ (3)

Multiply (2) by 2: $\qquad\qquad 6x + 4y + 16 = 0.$ (4)

Add: $\qquad\qquad\qquad\qquad 8x \qquad + 20 = 0.$ (5)

Solve (5) for x: $\qquad\qquad\qquad\qquad x = -\tfrac{5}{2}.$ (6)

Substitute $x = -\tfrac{5}{2}$ in (1): $\quad 2(-\tfrac{5}{2}) - 4y + 4 = 0.$ (7)

Solve (7) for y: $\qquad\qquad\qquad\qquad y = -\tfrac{1}{4}.$

Solution is $(-\tfrac{5}{2}, -\tfrac{1}{4})$.

Check. Substitute into **both** (1) and (2).

$2(-\tfrac{5}{2}) - 4(-\tfrac{1}{4}) + 4 = -5 + 1 + 4 = 0.$

$3(-\tfrac{5}{2}) + 2(-\tfrac{1}{4}) + 8 = -\tfrac{15}{2} - \tfrac{1}{2} + 8 = -\tfrac{16}{2} + 8 = 0.$

Illustration 2. Solve:

$$\begin{cases} 2x - 4y + 4 = 0, & (1) \\ 3x + 2y + 8 = 0. & (2) \end{cases}$$

Solution.

Solve (1) for $x = f(y)$: $x = 2y - 2.$ (3)

Substitute (3) in (2): $3(2y - 2) + 2y + 8 = 0.$ (4)

Solve (4) for y: $y = -\frac{1}{4}.$ (5)

Substitute $y = -\frac{1}{4}$ in (1): $2x - 4(-\frac{1}{4}) + 4 = 0.$ (6)

Solve (6) for x: $x = -\frac{5}{2}.$

Solution is $(-\frac{5}{2}, -\frac{1}{4})$.

In the first illustration we eliminated y by an addition after multiplying equation (2) through by a constant, 2, which does not change the equation. The system was thus reduced to one equation in one unknown which could easily be solved. The second unknown was then found by a substitution. Frequently it is necessary to multiply both equations to effect an elimination.

Illustration. Solve:

$$\begin{cases} 3x - 2y = 6, & (1) \\ 2x - 3y = -1. & (2) \end{cases}$$

Solution.

Multiply (1) by 2: $6x - 4y = 12.$ (3)

Multiply (2) by 3: $6x - 9y = -3.$ (4)

Subtract (4) from (3): $5y = 15.$ (5)

Solve (5) for y: $y = 3.$ (6)

Multiply (1) by 3: $9x - 6y = 18.$ (7)

Multiply (2) by 2: $4x - 6y = -2$ (8)

Subtract (8) from (7): $5x = 20.$ (9)

Solve (9) for x: $x = 4.$

Solution is (4,3).

Check. Substitute into (1) and (2).

$$3(4) - 2(3) = 12 - 6 = 6.$$
$$2(4) - 3(3) = 8 - 9 = -1.$$

After we have obtained one value, the second can always be obtained by substituting into the original equations. However, if an error has been made in finding the first value, an erroneous value for the second variable will be obtained. It is therefore always necessary to check in both equations.

Sometimes systems which appear to be nonlinear can be made linear by a change of variables.

Illustration. Solve:

$$\begin{cases} x - y + 12xy = 0, \\ 3x + y - 4xy = 0. \end{cases}$$

Solution. Divide both equations by xy:

$$\begin{cases} \dfrac{1}{y} - \dfrac{1}{x} + 12 = 0, \\[2mm] \dfrac{3}{y} + \dfrac{1}{x} - 4 = 0. \end{cases}$$

Replace $\dfrac{1}{x}$ by w and $\dfrac{1}{y}$ by v:

$$\begin{cases} v - w + 12 = 0, \\ 3v + w - 4 = 0. \end{cases}$$

Solving this system we obtain:

$$v = -2 \quad \text{and} \quad w = 10.$$

$$\therefore \quad x = \frac{1}{w} = \frac{1}{10} \quad \text{and} \quad y = \frac{1}{v} = -\frac{1}{2}.$$

Check.

$$\tfrac{1}{10} - (-\tfrac{1}{2}) + 12(\tfrac{1}{10})(-\tfrac{1}{2}) = \tfrac{6}{10} - \tfrac{6}{10} = 0.$$
$$3(\tfrac{1}{10}) + (-\tfrac{1}{2}) - 4(\tfrac{1}{10})(-\tfrac{1}{2}) = \tfrac{3}{10} - \tfrac{5}{10} + \tfrac{2}{10} = 0.$$

We should remember that when we divide by an unknown we may lose a solution. Thus, in the above example, we have $x = 0$, $y = 0$ as another solution. The problem properly belongs in the discussion of systems of quadratic equations.

61. Systems of Three Linear Equations in Three Unknowns. To solve a system of three linear equations in three unknowns, we eliminate the same unknown from any two pairs of the equations and solve the resulting two linear equations in two unknowns.* The value of the third unknown is then found by substituting into any of the original equations.

Illustration. Solve:

$$\begin{cases} 3x + 2y - z = 4, \\ x - y + 3z = 3, \\ 2x - 2y - z = -15. \end{cases}$$

Solution. Multiply the second equation by 2 and add to the first; then add the first and third equations.

$$5x + 5z = 10.$$
$$5x - 2z = -11.$$

Subtract: $\quad 7z = 21;$ $\quad z = 3.$

Substitute: $5x = -11 + 2(3) = -5;$ $\quad x = -1.$

---

* If this is not possible the equations may be studied by use of Sections 62 and 63.

Substitute: $2y = 4 + z - 3x = 10;$ $y = 5.$

Check. $3(-1) + 2(5) - (3) = -3 + 10 - 3 = 4.$
            $-1 - 5 + 3(3) = -6 + 9 = 3.$
            $2(-1) - 2(5) - (3) = -2 - 10 - 3 = -15.$

62. Determinants. We shall now introduce another mathematical symbol and define some of the operations with it. The general theory of this subject is beyond the scope of this book * and, in fact, forms a part of the study of advanced mathematics. The symbol:

$$\begin{vmatrix} a_1 & b_1 \\ a_2 & b_2 \end{vmatrix}$$

is called a **determinant.** More specifically, it is a *determinant of the second order.* It represents a number, and this number can be found by *expanding* the determinant; thus:

$$\begin{vmatrix} a_1 & b_1 \\ a_2 & b_2 \end{vmatrix} = a_1b_2 - a_2b_1$$

and the right side is said to be the expansion of the determinant. A determinant is made up of:

elements, the quantities a_1, a_2, b_1, b_2;
rows, the elements of a horizontal line;
columns, the elements of a vertical line;
principal diagonal, the elements a_1 and b_2, or the diagonal starting in the upper left-hand corner.

To evaluate a *second-order* determinant, multiply the elements of the principal diagonal and *subtract* from the product the product of elements of the other diagonal; schematically:

We shall use determinants to solve systems of linear equations. Consider the system:

$$a_1x + b_1y = c_1,$$
$$a_2x + b_2y = c_2.$$

---

* See R. A. Beaumont and R. W. Ball, *Introduction to Modern Algebra and Matrix Theory*, Rinehart, N.Y., 1954, Chapter I.

If we multiply the first equation by b_2, the second by $-b_1$, and add the resulting equations, we obtain:

$$(a_1b_2 - a_2b_1)x = b_2c_1 - b_1c_2.$$

If we now solve for x we write:

$$x = \frac{b_2c_1 - b_1c_2}{a_1b_2 - a_2b_1},$$

provided $a_1b_2 - a_2b_1 \neq 0$. Similarly, if we eliminate x between the given equations and solve for y, we have:

$$y = \frac{a_1c_2 - a_2c_1}{a_1b_2 - a_2b_1}.$$

We notice that both denominators are the same and that they can be represented by:

$$D = \begin{vmatrix} a_1 & b_1 \\ a_2 & b_2 \end{vmatrix}$$

This determinant, D, is called the determinant of the system. For the equations to be consistent, this determinant must *not be equal to zero*.

Let us study the numerators. We see that for the solution of x:

$$b_2c_1 - b_1c_2 = \begin{vmatrix} c_1 & b_1 \\ c_2 & b_2 \end{vmatrix} = N_x,$$

which is the determinant D with a_1 and a_2 (the coefficients of x) replaced by c_1 and c_2 (the constant terms), and that for the solution of y:

$$a_1c_2 - a_2c_1 = \begin{vmatrix} a_1 & c_1 \\ a_2 & c_2 \end{vmatrix} = N_y,$$

which is the determinant D with b_1 and b_2 (the coefficients of y) replaced by c_1 and c_2 (the constant terms).

Let us apply this knowledge to an example.

Illustration. Solve by the determinant formulas:

$$\begin{cases} 3x - 2y = 7, \\ x + 3y = 5. \end{cases}$$

Solution.

$$D = \begin{vmatrix} 3 & -2 \\ 1 & 3 \end{vmatrix} = (3)(3) - (-2)(1) = 9 + 2 = 11.$$

$$N_x = \begin{vmatrix} 7 & -2 \\ 5 & 3 \end{vmatrix} = (7)(3) - (-2)(5) = 21 + 10 = 31.$$

$$N_y = \begin{vmatrix} 3 & 7 \\ 1 & 5 \end{vmatrix} = (3)(5) - (7)(1) = 15 - 7 = 8.$$

$$\therefore \quad x = \tfrac{31}{11} \quad \text{and} \quad y = \tfrac{8}{11}.$$

Check the solution.

A natural question is, "What can we deduce if $D = 0$?" One answer is that the equations are either inconsistent or dependent. More precisely, it can be shown that:

(i) *If $D = 0$ and* either $N_x \neq 0$ or $N_y \neq 0$, *the equations are inconsistent.*

(ii) *If $D = 0$* and $N_x = N_y = 0$, *the equations are dependent.*

63. Third-Order Determinants. If a determinant has three rows and three columns, it is called a **third-order determinant.** The value of this determinant:

$$\begin{vmatrix} a_1 & b_1 & c_1 \\ a_2 & b_2 & c_2 \\ a_3 & b_3 & c_3 \end{vmatrix}$$

is defined to be:

$$a_1 b_2 c_3 + a_2 b_3 c_1 + a_3 b_1 c_2 - a_1 b_3 c_2 - a_3 b_2 c_1 - a_2 b_1 c_3.$$

A scheme for evaluating the third-order determinant is given by:

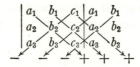

Rewrite the first two columns to the right of the determinant. Multiply the elements in each of the three diagonals running down from left to right; these products are prefixed with "+" signs and are the first three terms of the expansion. The last three terms are found by multiplying the elements in each of the three diagonals running down from right to left and prefixing all these products with "−" signs.

Illustration. Evaluate the determinant:

$$\begin{vmatrix} 3 & -1 & 7 \\ 2 & 3 & -1 \\ 1 & 5 & -3 \end{vmatrix}$$

Solution.

$$\begin{vmatrix} 3 & -1 & 7 \\ 2 & 3 & -1 \\ 1 & 5 & -3 \end{vmatrix} \begin{matrix} 3 & -1 \\ 2 & 3 \\ 1 & 5 \end{matrix} = \begin{aligned} & (3)(3)(-3) + (-1)(-1)(1) + (7)(2)(5) \\ & - (7)(3)(1) - (3)(-1)(5) - (-1)(2)(-3) \end{aligned}$$
$$= -27 + 1 + 70 - 21 + 15 - 6 = 32.$$

Third-order determinants are used to solve a system of three linear equations in three unknowns. Consider:

$$\begin{cases} a_1 x + b_1 y + c_1 z = d_1, \\ a_2 x + b_2 y + c_2 z = d_2, \\ a_3 x + b_3 y + c_3 z = d_3, \end{cases}$$

and let:

$$D = \begin{vmatrix} a_1 & b_1 & c_1 \\ a_2 & b_2 & c_2 \\ a_3 & b_3 & c_3 \end{vmatrix}$$

Then:

$$x = \frac{1}{D} \begin{vmatrix} d_1 & b_1 & c_1 \\ d_2 & b_2 & c_2 \\ d_3 & b_3 & c_3 \end{vmatrix}$$

$$y = \frac{1}{D} \begin{vmatrix} a_1 & d_1 & c_1 \\ a_2 & d_2 & c_2 \\ a_3 & d_3 & c_3 \end{vmatrix}$$

$$z = \frac{1}{D} \begin{vmatrix} a_1 & b_1 & d_1 \\ a_2 & b_2 & d_2 \\ a_3 & b_3 & d_3 \end{vmatrix}$$

if $D \neq 0$.

These formulas may be verified by solving the three general equations by the method of Section 61.

Illustration. Solve:

$$\begin{cases} 2x - 3y + 4z = 7, \\ x + y - z = 3, \\ 3x + 2y - 4z = 6. \end{cases}$$

Solution.

$$D = \begin{vmatrix} 2 & -3 & 4 \\ 1 & 1 & -1 \\ 3 & 2 & -4 \end{vmatrix} = -11; \qquad N_x = \begin{vmatrix} 7 & -3 & 4 \\ 3 & 1 & -1 \\ 6 & 2 & -4 \end{vmatrix} = -32;$$

$$N_y = \begin{vmatrix} 2 & 7 & 4 \\ 1 & 3 & -1 \\ 3 & 6 & -4 \end{vmatrix} = -17; \qquad N_z = \begin{vmatrix} 2 & -3 & 7 \\ 1 & 1 & 3 \\ 3 & 2 & 6 \end{vmatrix} = -16;$$

$\therefore \quad x = \frac{32}{11}, y = \frac{17}{11}, z = \frac{16}{11}.$

Check the solution.

64. Exercise V.

1. Draw the graphs of the following straight lines:

   (a) $3x + 2y - 6 = 0$.
   (b) $y = 4x + 2$.
   (c) $\frac{1}{3}x + \frac{1}{5}y = 1$.

2. Find the midpoint of the line joining the points $(7,-3)$ and $(-3,9)$.
3. Find the slopes of the three lines of Problem 1.
4. Find the equations of the following lines:

   (a) Parallel to the X-axis and through the point $(0,3)$.
   (b) With a slope of -3 and passing through $(2,2)$.
   (c) Through the points $(-1,3)$ and $(5,-2)$.
   (d) Having intercepts $a = 4$ and $b = 2$.
   (e) Passing through the point $(3,1)$ and parallel to the line

   $$y = 3x + 5.$$

   (f) Passing through the point $(-2,-3)$ and perpendicular to the line $2y - 6x + 7 = 0$.

5. Find the angle between the lines:

   $$2y - 6x + 7 = 0 \quad \text{and} \quad 6x - 3y = 5.$$

6. Find the distance from the point $(4,6)$ to the line $3x + 4y = 12$.
7. Solve the system:

   $$\begin{cases} 3x - 2y = 4, \\ 2x + 3y = 6. \end{cases}$$

8. Find the value of the following determinants:

   $$\text{(a)} \begin{vmatrix} 4 & -1 & 2 \\ 1 & 3 & -1 \\ 7 & 4 & 5 \end{vmatrix}; \quad \text{(b)} \begin{vmatrix} x & y & 1 \\ x_1 & y_1 & 1 \\ x_2 & y_2 & 1 \end{vmatrix}.$$

   Set (b) equal to zero and compare with Formula (3), Section 54.

9. Solve the system:

   $$\begin{cases} 4x - y + 2z = 1, \\ x + 3y - z = 2, \\ 7x + 4y + 5z = 3. \end{cases}$$

10. Find the value of:

(a) $\begin{vmatrix} -1 & 4 & 2 \\ 3 & 1 & -1 \\ 4 & 7 & 5 \end{vmatrix}$; (b) $\begin{vmatrix} 4 & -1 & 2 \\ 7 & 4 & 5 \\ 1 & 3 & -1 \end{vmatrix}$.

Compare both answers with Problem 8(a).

6

QUADRATIC EQUATIONS

65. The General Quadratic Equation. A quadratic equation is one in which the highest degree of the variables is two. The equation:

$$ax^2 + bx + c = 0, \quad a \neq 0,$$

where a, b, and c are constants, is called the **general quadratic equation in x.** If $b = 0$, the equation is called a *pure quadratic equation.* If $a = 0$, the equation reduces to the linear equation $bx + c = 0$. Any quadratic equation can be put into the general form by collecting the coefficients of each term in the unknown.

Illustration. Change the equation $3x^2 + rx + sx^2 = (2x + r)^2$ to the general form.

Solution. Multiply out the right side:

$$3x^2 + sx^2 + rx = 4x^2 + 4rx + r^2.$$

Collect terms:

$$(3 + s - 4)x^2 + (r - 4r)x - r^2 = 0,$$

or

$$(s - 1)x^2 - 3rx - r^2 = 0.$$

Thus: $\qquad a = s - 1, \quad b = -3r, \quad \text{and} \quad c = -r^2.$

66. The Quadratic Formula. By applying a process known as **completing the square** to the general quadratic equation, we can arrive at a formula which will give the values of the unknown in any quadratic equation. The process of completing the square is one which creates a perfect square on the left side of the equation. The procedure is as follows:

1. Transpose the constant term to the right side:

$$ax^2 + bx = -c.$$

2. Divide by the coefficient of x^2:

$$x^2 + \frac{b}{a}x = -\frac{c}{a}.$$

100

3. Take $\frac{1}{2}$ the coefficient of x, square it, and add to both sides:

$$x^2 + \frac{b}{a}x + \left(\frac{b}{2a}\right)^2 = \frac{b^2}{4a^2} - \frac{c}{a}.$$

4. Factor the left side and simplify the right side:

$$\left(x + \frac{b}{2a}\right)^2 = \frac{b^2 - 4ac}{4a^2}.$$

5. Extract the square root of each member:

$$x + \frac{b}{2a} = \frac{\pm\sqrt{b^2 - 4ac}}{2a}.$$

6. Solve for x:

$$x = \frac{-b \pm \sqrt{b^2 - 4ac}}{2a}.$$

This is known as the **quadratic formula.** It should be committed to memory.

The expression $b^2 - 4ac$ is known as the **discriminant** of the quadratic equation and can give us certain information about the roots of the equation. Thus:

*If the coefficients **a**, **b**, and **c** are real numbers, and:*

if	then the roots of $ax^2 + bx + c = 0$ are
$b^2 - 4ac > 0$	real and unequal
$b^2 - 4ac = 0$	real and equal
$b^2 - 4ac < 0$	not real

Furthermore, *if the roots are real and the coefficients are rational numbers, and:*

if	the roots are
$b^2 - 4ac$ is a perfect square	rational
$b^2 - 4ac$ is not a perfect square	irrational

67. Solution of Quadratic Equations.   There are three algebraic methods of finding the solutions of quadratic equations.

I. By factoring.   This applies the principle that if a product equals zero, one or more of its factors equal zero.

Illustration. Solve $8x^2 + 2x = 3$.
Solution.

Transpose: $\qquad\qquad\qquad\qquad 8x^2 + 2x - 3 = 0.$

Factor: $\qquad\qquad\qquad\qquad \underline{(4x + 3)(2x - 1) = 0.}$

Equate each factor to zero: $\qquad 4x + 3 = 0.$ | $2x - 1 = 0.$

Solve for x: $\qquad\qquad\qquad\qquad x = -\frac{3}{4}.$ | $x = \frac{1}{2}.$

Check. $\quad 8(\frac{9}{16}) + 2(-\frac{3}{4}) = \frac{9}{2} - \frac{3}{2} = \frac{6}{2} = 3.$

$\qquad\quad 8(\frac{1}{4}) + 2(\frac{1}{2}) = 2 + 1 = 3.$

II. By completing the square. Apply the process of completing the square to the given equation.

Illustration. Solve $2x^2 - 3x = 9$.
Solution.

Divide by 2: $\qquad\qquad\qquad\qquad x^2 - \frac{3}{2}x = \frac{9}{2}.$

Complete the square: $\qquad x^2 - \frac{3}{2}x + (\frac{3}{4})^2 = \frac{9}{2} + \frac{9}{16}.$

$\qquad\qquad\qquad\qquad\qquad\quad (x - \frac{3}{4})^2 = \frac{81}{16}.$

Extract square root: $\qquad\qquad\qquad x - \frac{3}{4} = \pm\frac{9}{4}.$

Solve for x: $\qquad\qquad\qquad\qquad x = \frac{3}{4} \pm \frac{9}{4}.$

$\qquad\qquad\qquad\qquad x = \frac{12}{4} = 3; \; x = -\frac{6}{4} = -\frac{3}{2}.$

Check. $\qquad 2(9) - 3(3) = 18 - 9 = 9.$

$\qquad\quad 2(\frac{9}{4}) - 3(-\frac{3}{2}) = \frac{9}{2} + \frac{9}{2} = 9.$

III. By the quadratic formula. Substitute the given coefficients into the quadratic formula and simplify.

Illustration. Solve $2x^2 + 3x + 4 = 0$.
Solution. $\qquad\qquad a = 2, \quad b = 3, \quad$ and $\quad c = 4.$

$$x = \frac{-b \pm \sqrt{b^2 - 4ac}}{2a}$$

$$= \frac{-3 \pm \sqrt{9 - 4(2)(4)}}{2(2)}$$

$$= \frac{-3 \pm \sqrt{9 - 32}}{4}$$

$$= \frac{-3 \pm \sqrt{-23}}{4} = \frac{1}{4}(-3 \pm i\sqrt{23}).$$

It should be noted that it may not be easy to factor the quadratic, and that completing the square is deriving the formula for a particular equation.

If we let:

$$r_1 = \frac{-b + \sqrt{b^2 - 4ac}}{2a} \quad \text{and} \quad r_2 = \frac{-b - \sqrt{b^2 - 4ac}}{2a}$$

we can easily obtain the following two expressions:

$$r_1 + r_2 = -\frac{b}{a} \quad \text{and} \quad r_1 \cdot r_2 = \frac{c}{a}.$$

These expressions form easy checks on our work and can be used to find the other root if one is known.

68. The Graph of a Quadratic Function. A function of the form:

$$f(x) = ax^2 + bx + c, \quad a \neq 0,$$

is called a *quadratic function*, and its graph is a *parabola*. To obtain the graph we plot a series of points and join them with a smooth curve.

Illustration 1. Find the graph of:

$$f(x) = 8x^2 + 2x - 3.$$

Solution. Calculate a table of values.

x	-1	$-\frac{1}{2}$	0	$\frac{1}{2}$	1
$f(x)$	3	-2	-3	0	7

Plot the points as shown in Fig. 38.

Illustration 2. Find the graph of:

$$f(x) = -4x^2 + 4x + 3.$$

Solution. Calculate a table of values.

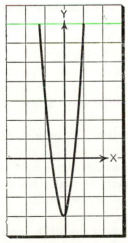

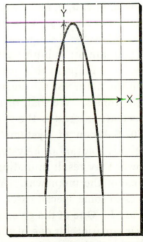

FIG. 38   FIG. 39

x	-1	$-\frac{1}{2}$	0	$\frac{1}{2}$	1	2
$f(x)$	-5	0	3	4	3	-5

Plot the points as shown in Fig. 39, p. 103.

We noticed in the two illustrations that if $a > 0$, then the curve opens *upward* and if $a < 0$, the curve opens *downward*.

The real zeros of $f(x)$ are the real roots of $f(x) = 0$ and can be found approximately by reading the points where the graph crosses the X-axis. The graph will also tell us if there are two real roots of the equation $f(x) = 0$. In fact we have:

if	the roots of $f(x) = 0$ are
the graph of $f(x)$ crosses the X-axis in two points	real and unequal
the graph of $f(x)$ touches the X-axis in only one point	real and equal
the graph of $f(x)$ does not touch the X-axis at all	not real

69. A Quadratic in Two Variables. The equation:

$$ax^2 + bxy + cy^2 + dx + ey + f = 0,$$

where a, b, c, d, e, and f are constants, is called the **general quadratic equation in x and y**. It is further assumed that a, b, and c are not *all* zero; otherwise the equation would reduce to a linear equation.

There are an infinite number of points (x,y) satisfying this equation. The graph of this quadratic equation is a curve and, depending upon the nature of the equation, it is one of a group of curves known as the **conics**. We shall discuss these conics in the next chapter. Of immediate interest is the system of two such equations.

70. Solution of Systems Involving Quadratics. In general, the algebraic solution of a system of two quadratic equations in two unknowns may result in finding the solution of a fourth-degree equation in one of the unknowns. At this time, however, we shall limit ourselves to some special cases which may be handled more easily.

Case I. **One of the given equations is linear.** In this case solve the *linear* equation for one of the unknowns in terms of the other, substitute into the quadratic equation, thus reducing it to a quadratic equation in one unknown, and solve the resulting quadratic equation. Finally, substitute the results into the linear equation for the corresponding values of the other unknown.

Illustration. Solve:

$$\begin{cases} x^2 - 4x + y^2 - 6y + 4 = 0, & (1) \\ x + y = 8. & (2) \end{cases}$$

Solution.

Solve (2) for y:	$y = 8 - x$.	(3)
Substitute (3) in (1):	$x^2 - 4x + (8 - x)^2 - 6(8 - x) + 4 = 0$.	(4)
Simplify (4):	$x^2 - 4x + 64 - 16x + x^2 - 48 + 6x + 4 = 0$.	
	$2x^2 - 14x + 20 = 0$.	
	$x^2 - 7x + 10 = 0$.	(5)
Solve (5):	$(x - 5)(x - 2) = 0$.	
	$x = 5$ and $x = 2$.	
Substitute in (3):	$y = 8 - 5 = 3$ and $y = 8 - 2 = 6$.	

Answers: $x = 5, y = 3$ and $x = 2, y = 6$.

It is important that the solutions be properly paired. Thus, in the example, $x = 5$ and $y = 6$ would *not* be a solution. The answers are frequently written in the coordinate notation for a point (x_1, y_1). In the illustration: $(5,3)$ and $(2,6)$.

Case II. **One equation of the form $bxy + f = 0$; the other of the form** $ax^2 + cy^2 + f = 0$. This system is treated in the same way as Case I. The equation $bxy + f = 0$ is solved for x or y and then substituted into $ax^2 + cy^2 + f = 0$. The resulting equation is a fourth-degree equation which is quadratic in the square of one unknown and can be solved by familiar methods.

Illustration. Solve:

$$\begin{cases} xy = 4, & (1) \\ x^2 + 4y^2 = 20. & (2) \end{cases}$$

Solution.

Solve (1) for y:	$y = \dfrac{4}{x}.$	(3)
Substitute (3) in (2):	$x^2 + 4\left(\dfrac{16}{x^2}\right) = 20.$	(4)
Simplify (4):	$x^4 + 64 = 20x^2.$	
	$x^4 - 20x^2 + 64 = 0.$	(5)
Solve (5) for x^2:	$(x^2 - 16)(x^2 - 4) = 0.$	
	$x^2 = 16$ and $x^2 = 4.$	
Solve for x:	$x = \pm 4$ and $x = \pm 2.$	

Substitute into (3): $\quad y = \dfrac{4}{\pm 4} = \pm 1 \;$ and $\; y = \dfrac{4}{\pm 2} = \pm 2.$

Answers: $(4,1)$; $(-4,-1)$; $(2,2)$; $(-2,-2)$.

Case III. **Both equations of the form** $ax^2 + cy^2 + f = 0$. In this case both equations are linear in x^2 and y^2, and the methods of solving systems of linear equations can be applied.

Illustration. Solve:

$$\begin{cases} 4x^2 + 5y^2 = 16, & (1) \\ 13y^2 - 5x^2 = 57. & (2) \end{cases}$$

Solution.

Multiply (1) by 5: $\qquad\qquad\qquad 20x^2 + 25y^2 = 80.$ $\qquad$ (3)

Multiply (2) by 4: $\qquad\qquad\quad -20x^2 + 52y^2 = 228.$ $\qquad$ (4)

Add (3) and (4): $\qquad\qquad\qquad\qquad\quad 77y^2 = 308.$

Solve for y: $\qquad\qquad\qquad\qquad\qquad\; y^2 = 4.$

$$y = \pm 2.$$

Substitute in (1): $\quad$
$4x^2 + 5(2)^2 = 16.$	$4x^2 + 5(-2)^2 = 16.$
$x^2 = -1.$	$x^2 = -1.$
$x = \pm i.$	$x = \pm i.$

Answers: $(i,2)$; $(-i,2)$; $(i,-2)$; $(-i,-2)$.

Case IV. **One equation of the form** $ax^2 + cy^2 + f = 0$; **the other of the form** $ax^2 + ey + f = 0$ **or** $cy^2 + dx + f = 0$. To solve a system of this type, eliminate the variable which is quadratic in both equations and solve the resulting quadratic equation in one unknown. Substitute to get the value of the second unknown.

Illustration. Solve:

$$\begin{cases} x^2 + 4y^2 - 25 = 0, & (1) \\ x^2 - 2y - 5 = 0. & (2) \end{cases}$$

Solution.

Subtract (2) from (1): $\qquad\qquad 4y^2 + 2y - 20 = 0.$ $\qquad$ (3)

Solve (3) for y: $\qquad\qquad\quad \underline{(2y + 5)(y - 2) = 0.}$

Substitute in (1): $\quad$
$y = -\frac{5}{2}.$	$y = 2.$
$x^2 = 25 - 4(\frac{25}{4})$	$x^2 = 25 - (4)(4)$
$= 0.$	$= 9.$

Solve for x: $\qquad\qquad\quad$
$x = 0.$	$x = \pm 3.$

Answers: $(0,-\frac{5}{2})$; $(3,2)$; $(-3,2)$.

Case V. **Both equations of the form** $ax^2 + bxy + cy^2 + f = 0$. This system is solved by eliminating the constant term, f. The resulting equation is factored into two linear factors, and each factor is considered with one of the given equations. The two systems thus formed are solved by the method of Case I.

Illustration. Solve:

$$\begin{cases} 3x^2 + xy + 2y^2 - 6 = 0, & (1) \\ 3x^2 + xy + 4y^2 - 9 = 0. & (2) \end{cases}$$

Solution.

Multiply (1) by 3:	$9x^2 + 3xy + 6y^2 - 18 = 0.$	(3)
Multiply (2) by 2:	$6x^2 + 2xy + 8y^2 - 18 = 0.$	(4)
Subtract (4) from (3):	$3x^2 + \ xy - 2y^2 \qquad = 0.$	(5)
Factor (5):	$(3x - 2y)(x + y) = 0.$	

	$3x = 2y.$	$x + y = 0.$ (6)
	$x = \tfrac{2}{3}y.$	$x = -y.$
Substitute in (1):	$3(\tfrac{2}{3}y)^2+(\tfrac{2}{3}y)y+2y^2-6 = 0.$	$3(-y)^2+(-y)y+2y^2-6 = 0.$
	$\tfrac{4}{3}y^2 + \tfrac{2}{3}y^2 + 2y^2 - 6 = 0.$	$3y^2 - y^2 + 2y^2 - 6 = 0.$
	$\tfrac{12}{3}y^2 = 6.$	$4y^2 = 6.$
	$y^2 = \tfrac{6}{4}.$	$y^2 = \tfrac{6}{4}.$
	$y = \pm\tfrac{1}{2}\sqrt{6}.$	$y = \pm\tfrac{1}{2}\sqrt{6}.$
Substitute in (6):	$x = \pm\tfrac{2}{3}(\pm\tfrac{1}{2}\sqrt{6}).$	$x = \mp\tfrac{1}{2}\sqrt{6}.$

Answers: $(\tfrac{1}{3}\sqrt{6}, \tfrac{1}{2}\sqrt{6})$; $(-\tfrac{1}{3}\sqrt{6}, -\tfrac{1}{2}\sqrt{6})$; $(-\tfrac{1}{2}\sqrt{6}, \tfrac{1}{2}\sqrt{6})$; $(\tfrac{1}{2}\sqrt{6}, -\tfrac{1}{2}\sqrt{6})$.

Case VI. **Both equations of the form:**

$$A(x^2 + y^2) + Bxy + D(x + y) + F = 0.$$

An equation of this form is said to be *symmetrical* in x and y because it is unchanged if we interchange x and y. The system is solved by letting:

$$x = u + v \quad \text{and} \quad y = u - v,$$

so that:
$$x^2 + y^2 = 2(u^2 + v^2);$$
$$xy = u^2 - v^2;$$
$$x + y = 2u.$$

After substituting into the given equations, each reduces to the form:

$$2Au^2 + 2Av^2 + Bu^2 - Bv^2 + 2Du + F = 0,$$

or

(VI) $\qquad (2A + B)u^2 + (2A - B)v^2 + 2Du + F = 0.$

We see that v appears only as a squared term and thus can easily be eliminated, leaving a quadratic equation in only one unknown, u.

Illustration. Solve:

$$\begin{cases} x^2 + 3xy + y^2 + 2x + 2y - 5 = 0, & (1) \\ 3x^2 - 2xy + 3y^2 - 5x - 5y - 26 = 0. & (2) \end{cases}$$

Solution. Let $x = u + v$ and $y = u - v$; then by formula (VI) above:

$$(2 + 3)u^2 + (2 - 3)v^2 + 2(2)u - 5 = 0,$$
$$(6 - 2)u^2 + (6 + 2)v^2 + 2(-5)u - 26 = 0,$$

or

$$5u^2 - v^2 + 4u - 5 = 0, \tag{3}$$
$$4u^2 + 8v^2 - 10u - 26 = 0. \tag{4}$$

Add 8 times (3) to (4):

$$44u^2 + 22u - 66 = 0,$$
or $\qquad\qquad 2u^2 + u - 3 = 0. \tag{5}$

Factor (5): $\qquad (2u + 3)(u - 1) = 0.$

Solve for u: $u = -\frac{3}{2}$ and $u = 1$.

Substitute these values for u into (3) and solve for v:

$$5(\tfrac{9}{4}) - v^2 + 4(-\tfrac{3}{2}) - 5 = 0,$$
or $\qquad\qquad v^2 = \tfrac{1}{4}$ and $v = \pm\tfrac{1}{2}.$

Also:

$$5(1) - v^2 + 4(1) - 5 = 0,$$
or $\qquad\qquad v^2 = 4$ and $v = \pm 2.$

Thus:

u	$-\frac{3}{2}$	$-\frac{3}{2}$	1	1
v	$\frac{1}{2}$	$-\frac{1}{2}$	2	-2
x	-1	-2	3	-1
y	-2	-1	-1	3

Answers: $(-1,-2)$; $(-2,-1)$; $(3,-1)$; $(-1,3)$.

It is not recommended that formula (VI) above be memorized but that the student substitute $x = u + v$ and $y = u - v$ directly into the given equations.

The method of substituting new variables can also be used for the equations of Case V by letting $y = mx$, eliminating x by solving both equations for x^2, and solving the resulting quadratic equation for m.

Illustration. Solve:

$$\begin{cases} x^2 + 3xy = 5, & (1) \\ x^2 - y^2 = 3. & (2) \end{cases}$$

Solution. Let $y = mx$; then:

$$\begin{cases} x^2 + 3mx^2 = 5, \\ x^2 - m^2x^2 = 3, \end{cases}$$

or

$$x^2 = \frac{5}{1 + 3m} \quad \text{and} \quad x^2 = \frac{3}{1 - m^2}, \tag{3}$$

so that:

$$\frac{5}{1 + 3m} = \frac{3}{1 - m^2},$$

or
$$5m^2 + 9m - 2 = 0.$$

Solving for m:

$$m = \tfrac{1}{5} \quad \text{and} \quad m = -2.$$

Substitute in (3):

$$x^2 = \tfrac{25}{8} \quad \text{and} \quad x^2 = -1,$$

so that:

$$x = \pm\tfrac{5}{4}\sqrt{2} \quad \text{and} \quad x = \pm i.$$

We now have $y = mx$ or:

$$y = \tfrac{1}{5}(\pm\tfrac{5}{4}\sqrt{2}) = \pm\tfrac{1}{4}\sqrt{2};$$
$$y = -2(\pm i) = \mp 2i.$$

Answers: $(\tfrac{5}{4}\sqrt{2}, \tfrac{1}{4}\sqrt{2})$; $(-\tfrac{5}{4}\sqrt{2}, -\tfrac{1}{4}\sqrt{2})$; $(i, -2i)$; $(-i, 2i)$.

It should be remembered that these six cases will not solve all systems of quadratic equations. There is no general procedure, except possibly the method of solving one equation for one unknown and substituting into the other equation. This will yield an equation in one unknown which, in general, will be of the fourth degree and may be difficult to solve. To find out whether or not there are any real solutions we may graph the two equations and see if their graphs intersect. The graphs of such equations are more easily obtained after we study the conics in the next chapter.

71. The Circle. Consider the special case of a quadratic in two variables in which $a = c$ and $b = 0$:

$$ax^2 + ay^2 + dx + ey + f = 0. \tag{1}$$

The case is especially interesting from the geometric point of view since it defines a circle. Let us, however, start with the precise definition of a circle.

Definitions.

A **circle** is the locus of a point that moves at a constant distance from a fixed point.

The fixed point is called the **center.**

The constant distance is called the **radius.**

The segment of a straight line lying inside the circle and passing through the center is called the **diameter.**

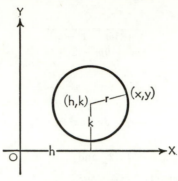

Fig. 40

Let the center be at (h,k) and let the fixed distance be r. Then the distance from any point (x,y), on the circumference of the circle, to the center is given by (see Section 20):

$$\sqrt{(x - h)^2 + (y - k)^2} = r,$$

or, squaring both sides:

(2) $$(x - h)^2 + (y - k)^2 = r^2.$$

If the center is at the origin, then $h = 0 = k$ and the equation reduces to:

(3) $$x^2 + y^2 = r^2.$$

We thus have three standard forms for the equation of a circle, and of these equation (2) gives the characteristics most directly. To find the center and radius of the circle given by an equation of type (1) we use the method of completing the square.

Illustration. Find the center and radius of the circle:

$$x^2 + y^2 - 6x + 4y + 4 = 0.$$

Solution. Transpose the constant term and complete the square in x and y:

$$x^2 - 6x + 9 + y^2 + 4y + 4 = -4 + 9 + 4,$$

or

$$(x - 3)^2 + (y + 2)^2 = 9.$$

The center is the point $C(3,-2)$, and the radius is 3.

To plot a circle is a very simple matter, since after finding the center and radius the circle may be drawn with compasses.

72. Conditions Determining a Circle. Three conditions are necessary and sufficient to determine a circle. This may be seen from the equation of a circle:

(1) $$(x - h)^2 + (y - k)^2 = r^2,$$

in that if we determine h, k, and r the circle is determined.* Also the equation:

$$ax^2 + ay^2 + dx + ey + f = 0$$

may be written by dividing by a:

(2) $$x^2 + y^2 + Dx + Ey + F = 0,$$

and the circle is determined after we know D, E, and F. The general theorem is:

The number of conditions required to determine a curve is equal to the number of independent constants in the equation of the curve.

To determine the circle from three conditions we may use either equation (1) or (2).

Illustration. Find the equation of the circle through the points $P_1(1,1)$, $P_2(3,-2)$, and $P_3(5,6)$.

Solution. Substitute the coordinates of the given points into equation (2) to obtain:

$$\begin{cases} 2 + D + E + F = 0, \\ 13 + 3D - 2E + F = 0, \\ 61 + 5D + 6E + F = 0. \end{cases}$$

The system of three linear equations in three unknowns may be solved by elimination. The solutions are:

$$D = -\tfrac{116}{11}, \quad E = -\tfrac{37}{11}, \quad \text{and} \quad F = \tfrac{131}{11}.$$

The desired equation is:

$$11x^2 + 11y^2 - 116x - 37y + 131 = 0.$$

---

* In general, the coordinates of a point yield only one *condition*. However, h and k are the coordinates of a special point (the center) and as such determine two of the independent constants in the equation of a circle.

Illustration. Find the equation of the circle tangent to the line $2y = 3x$ at $(2,3)$ and passing through $(4,5)$.

Solution. From geometry we know that the radius is perpendicular to the tangent at the point of tangency. The slope of a line perpendicular to $2y = 3x$ is $-\frac{2}{3}$. The equation of a line with this slope and passing through the point $(2,3)$ is given by $3y + 2x = 13$ since:

$$y = mx + b$$

or

$$y = -\tfrac{2}{3}x + b$$

and $\qquad\qquad b = (3) + 2(\tfrac{2}{3}) = \tfrac{13}{3}.$

Since the center (h,k) must lie on this line we have:

$$3k + 2h = 13,$$

and since the circle passes through the points $(2,3)$ and $(4,5)$ we have:

$$(2 - h)^2 + (3 - k)^2 = r^2,$$
$$(4 - h)^2 + (5 - k)^2 = r^2.$$

Eliminate r by a subtraction to obtain the two equations:

$$\begin{cases} h + k = 7, \\ 3k + 2h = 13. \end{cases}$$

Solve for h and k: $h = 8$, $k = -1$.
Solve for r^2:

$$r^2 = (2 - 8)^2 + (3 + 1)^2 = 36 + 16 = 52.$$

The equation of the circle is:

$$(x - 8)^2 + (y + 1)^2 = 52.$$

73. Radical Axis. Since the equation of a circle can always be written in the form:

$$x^2 + y^2 + Dx + Ey + F = 0,$$

it may be possible to obtain a first-degree equation by subtracting the equation of one circle from that of another.

$$x^2 + y^2 + D_1x + E_1y + F_1 = 0$$
$$x^2 + y^2 + D_2x + E_2y + F_2 = 0$$

(1) $\qquad \overline{(D_1 - D_2)x + (E_1 - E_2)y + (F_1 - F_2) = 0.}$

This line is called the **radical axis**. If the circles intersect in two points this line is a **common chord**. If the circles are tangent to each other this line is the **common tangent**. Concentric circles do not have a radical axis since in this case $D_1 = D_2$ and $E_1 = E_2$.

Since the radical axis must pass through the points (or point) of intersection of the two circles, it can aid us in solving systems of quadratic equations in which the equations are circles.

Illustration. Find the equation of the common chord, the points of intersection, and the length of the common chord of the circles:

$$\begin{cases} 2x^2 + 2y^2 - 12x + 8y + 13 = 0, \\ 2x^2 + 2y^2 - 2x - 2y - 7 = 0. \end{cases}$$

Solution. Equation of common chord is:

$$-10x + 10y + 20 = 0$$

or

$$y - x + 2 = 0.$$

Solve for y:

$$y = x - 2.$$

Substitute into the second circle:

$$2[x^2 + x^2 - 4x + 4 - x - x + 2] - 7 = 0,$$
$$4x^2 - 12x + 5 = 0,$$
$$(2x - 5)(2x - 1) = 0,$$
$$x = \tfrac{5}{2}, \quad x = \tfrac{1}{2}.$$

Then: $y = \tfrac{5}{2} - 2 = \tfrac{1}{2}$ and $y = \tfrac{1}{2} - 2 = -\tfrac{3}{2}.$

The points of intersection are:

$$(\tfrac{5}{2}, \tfrac{1}{2}) \quad \text{and} \quad (\tfrac{1}{2}, -\tfrac{3}{2}).$$

The length of the common chord is:

$$d = \sqrt{(\tfrac{5}{2} - \tfrac{1}{2})^2 + (\tfrac{1}{2} + \tfrac{3}{2})^2} = \sqrt{4 + 4} = 2\sqrt{2}.$$

74. Exercise VI.

1. Solve the following quadratic equations by the method of factoring and the method of completing the square:

   (a) $6x^2 + x - 2 = 0.$

   (b) $8x^2 + 26x + 21 = 0.$

   (c) $9x^2 + 12x + 4 = 0.$

2. Solve by the quadratic formula:

   (a) $x^2 + x + 1 = 0.$

   (b) $4x^2 - 4x - 11 = 0.$

   (c) $16x^2 - 24x + 13 = 0.$

3. Find the graphs of the quadratic functions obtained by letting y equal the left side of 1(a), 2(a), and 2(b).

4. Determine the quadratic equation which has the sum of its roots equal to 9 and the product of its roots equal to 14. [Hint: see Section 67.]

5. Solve the equation obtained in Problem 4.

6. Solve the following systems of quadratic equations:

(a) $\begin{cases} 2x^2 + 9x - 9y - 18 = 0, \\ 3y - 5x + 6 = 0. \end{cases}$

(b) $\begin{cases} 2xy + 5 = 0, \\ 4x^2 + 4y^2 - 29 = 0. \end{cases}$

(c) $\begin{cases} 9x^2 + 16y^2 - 100 = 0, \\ y^2 - 4 = 0. \end{cases}$

(d) $\begin{cases} 3x^2 + 4y^2 - 5 = 0, \\ 4x^2 - 3y^2 + 7 = 0. \end{cases}$

(e) $\begin{cases} 3x^2 - 4xy + 2y^2 + 3 = 0, \\ x^2 + 3xy - 4y^2 - 2 = 0. \end{cases}$

(f) $\begin{cases} 2x^2 + 2y^2 + 6xy + x + y - 12 = 0, \\ 3x^2 + 3y^2 - xy - 4x - 4y + 3 = 0. \end{cases}$

7. Find the center and radius of the following circles. Draw each circle.

(a) $x^2 + y^2 - 6x - 14y + 22 = 0.$
(b) $16x^2 + 16y^2 + 80x + 51 = 0.$
(c) $225x^2 + 225y^2 + 150x + 180y = 189.$

8. Find the equations of the following circles.

(a) Through the points $(3,0)$, $(0,3)$, and $(-3,0)$.
(b) Through the points $(3,-2)$, $(5,3)$, and $(-1,9)$.
(c) Tangent to the line $y - x = 2$ at $(1,3)$ and through the point $(2,3)$.

9. Find the equation of the common chord, the points of intersection, and the length of the common chord of the circles:

$$\begin{cases} x^2 + y^2 - 6x + 4y = 3, \\ x^2 + y^2 + 2x - 4y = 11. \end{cases}$$

10. Solve:

$$\begin{cases} x^2 - 2x + y^2 - 2y = 47, \\ y^2 - 2y + x^2 - 6x = 15. \end{cases}$$

7

THE CONIC SECTIONS

75. Definitions. To continue our study of a quadratic equation in two variables, we shall now consider the conic sections, so called because they can be obtained by cutting a cone with a plane. There is more than one way to define a conic section.

I. A **conic section** is a curve whose equation is of the second degree.

This definition has the advantage of including the *degenerate* conics, i.e., those which reduce to points or lines.

II. A **conic section** is the path of a point which moves so that the *ratio* of its distance from a *fixed point* to its distance from a *fixed line* is *constant*.

This definition makes it easy to classify the conics. Other definitions are somewhat specialized and will be considered individually.

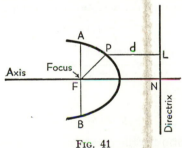

Fig. 41

In definition II, the fixed point is called the **focus**; the fixed line is called the **directrix**; and the constant ratio, the **eccentricity**. Fig. 41 shows the terminology. Here we note that the ratio defining the conic is:

$$\frac{PF}{PL} = \frac{n}{d} = e,$$

where e denotes the eccentricity. The line through the focus perpendicular to the directrix is an *axis of symmetry* * for the curve. The line through the focus parallel to the directrix intersects the curve in two points, A and B, and the line segment AB is called the **latus rectum.**

The conic sections are divided into three classes as follows:

> if $e < 1$, the conic is an *ellipse;*
> if $e = 1$, the conic is a *parabola;*
> if $e > 1$, the conic is a *hyperbola.*

If $e = 0$ the definition does not have a meaning, but as e *approaches* zero the ellipse approaches a circle, and the circle becomes the limiting case as e becomes zero.

76. The Parabola. There are two definitions of the parabola.

I. *The locus of points which are equidistant from a fixed point and a fixed line is called a* **parabola.**

II. *The conic section for which $e = 1$ is called a* **parabola.**

The axis of symmetry is the axis of this curve, and the point where the axis intersects the curve is called the **vertex** of the parabola.

Let us take the vertex of a parabola as the origin and the focus at $(a,0)$ where a is positive.** The axis of the curve is the X-axis; the directrix is the line $x = -a$; let $P(x,y)$ be any point on the curve. We have the configuration of Fig. 42. The distance from the point to the directrix is $a + x$, and the distance between P and F is:

$$\sqrt{(x - a)^2 + y^2}.$$

The definition of the parabola states:

$$\sqrt{(x - a)^2 + y^2} = a + x,$$

which can be simplified to:

$$(1) \qquad\qquad y^2 = 4ax.$$

This is a standard form for the parabola. In this case the curve lies entirely to the right of the Y-axis.

To obtain other standard forms let us use the property of formula (1) and consider the vertex at (h,k), the axis parallel to the X-axis,

---

* If a line is the perpendicular bisector of a segment such as AB it is called an **axis of symmetry,** and the points are said to be **symmetric** with respect to the line.

** In all cases for the parabola we shall denote the distance from the vertex to the focus by a.

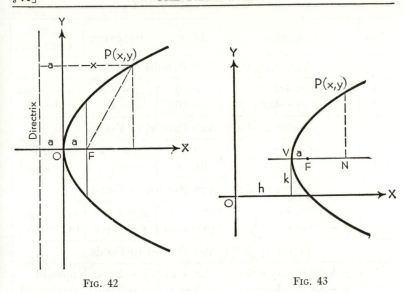

FIG. 42   FIG. 43

and the focus at a distance a to the right of the vertex, $F(h + a,k)$. We have the configuration of Fig. 43. By formula (1):

$$\overline{PN}^2 = 4(VF)(VN).$$

Now:
$$PN = y - k,$$
$$VF = a,$$
$$VN = x - h,$$

so that:

(2) $(y - k)^2 = 4a(x - h).$

The coordinates of the end points of the latus rectum are $(h + a,k + 2a)$ and $(h + a,k - 2a)$, obtained by letting $x = h + a$ in (2) and solving for y. The length of the latus rectum is obtained by the distance formula:

$$\sqrt{(h + a - h - a)^2 + (k + 2a - k + 2a)^2} = 4a.$$

We could choose the axis parallel to the Y-axis and let the curve open upward or downward. We could also have the curve parallel to the X-axis open to the left. There are four such combinations. We present all the standard forms of a parabola in the following table, and it is recommended that the student sketch each of the curves.

No.	Equation	Focus	Directrix	Curve Opening
Vertex at (0,0); Axis Parallel to X-axis				
1	$y^2 = 4ax$	$(a,0)$	$x = -a$	to the right
2	$y^2 = -4ax$	$(-a,0)$	$x = a$	to the left
Vertex at (0,0); Axis Parallel to Y-axis				
3	$x^2 = 4ay$	$(0,a)$	$y = -a$	upward
4	$x^2 = -4ay$	$(0,-a)$	$y = a$	downward
Vertex at (h,k); Axis Parallel to X-axis				
5	$(y - k)^2 = 4a(x - h)$	$(h + a,k)$	$x = h - a$	to the right
6	$(y - k)^2 = -4a(x - h)$	$(h - a,k)$	$x = h + a$	to the left
Vertex at (h,k); Axis Parallel to Y-axis				
7	$(x - h)^2 = 4a(y - k)$	$(h,k + a)$	$y = k - a$	upward
8	$(x - h)^2 = -4a(y - k)$	$(h,k - a)$	$y = k + a$	downward
Length of latus rectum: $4a$.				

The quadratic function:

$$y = Ax^2 + Bx + C, \quad (A \neq 0),$$

is a parabola with vertical axis. This can be shown by changing the equation into a standard form. Divide through by A and rewrite:

$$x^2 + \frac{B}{A} x = \frac{1}{A} (y - C).$$

Complete the square on the left member:

$$\left(x + \frac{B}{2A}\right)^2 = \frac{1}{A} \left(y - C + \frac{B^2}{4A}\right)$$

$$= \frac{1}{A} \left[y - \left(\frac{4AC - B^2}{4A}\right)\right],$$

which is either standard form 7 or 8 with:

$$h = -\frac{B}{2A}, \quad k = \frac{4AC - B^2}{4A}, \quad \text{and} \quad 4a = \frac{1}{A}.$$

Illustration. Find the vertex and focus of:

$$y = 2x^2 - 3x + 5.$$

Solution.

Divide by 2 and transpose: $x^2 - \frac{3}{2}x = \frac{1}{2}(y - 5)$.

Complete square: $(x - \frac{3}{4})^2 = \frac{1}{2}(y - 5 + \frac{9}{8})$.

Simplify: $(x - \frac{3}{4})^2 = \frac{1}{2}(y - \frac{31}{8})$.

Vertex: $(\frac{3}{4}, \frac{31}{8})$; $a = \frac{1}{8}$; Focus: $(\frac{3}{4}, 4)$.

The motion of a body moving in a vertical line under the influence of the earth's attraction only is approximated by the equation:

$$d = V_0 t - 16t^2,$$

where V_0 is the initial velocity; d, the distance; t, the time; both V_0 and d considered to be positive *upward*. This quadratic function is a parabola, and the height to which the body moves is at the vertex.

Illustration. A ball is thrown upward with a velocity of 48 feet per second. Find how far and for how long a time it will rise.

Solution. Since $V_0 = 48$ we have:

$$d = 48t - 16t^2$$

or

$$16(t^2 - 3t) = -d.$$

Complete the square:

$$(t - \frac{3}{2})^2 = -\frac{1}{16}(d - 36).$$

Vertex: $(\frac{3}{2}, 36)$. The ball will rise for $\frac{3}{2}$ seconds to a height of 36 feet above the starting point.

Let us consider the general equation of the second degree:

$$Ax^2 + Bxy + Cy^2 + Dx + Ey + F = 0.$$

If $B = 0$ and either $A = 0$ or $C = 0$, then the equation represents a parabola.

Illustration. Reduce the equation:

$$4y^2 - 16x - 20y - 7 = 0$$

to a standard form and trace the curve.

Solution. Divide by 4:

$$y^2 - 4x - 5y - \frac{7}{4} = 0.$$

Transpose the term in x and the constant, and complete the square in y:

$$y^2 - 5y + \frac{25}{4} = 4x + \frac{7}{4} + \frac{25}{4}$$

or

$$(y - \frac{5}{2})^2 = 4(x + 2).$$

The vertex is at $(-2, \frac{5}{2})$, $a = 1$, and the curve opens to the right. See Fig. 44.

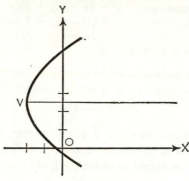

Fig. 44

Illustration. Find the equation of a parabola with its axis parallel to OY and passing through (1,1), (0,5), and (2,5).

Solution. Since the axis is parallel to OY, we have $B = C = 0$. Let *
$A = 1$; then the equation has the form:

$$x^2 + Dx + Ey + F = 0.$$

Substitute into this equation the coordinates of the three points, resulting in three linear equations in three unknowns:

$$\begin{cases} 1 + D + E + F = 0, \\ 5E + F = 0, \\ 4 + 2D + 5E + F = 0. \end{cases}$$

The solution of this system is:

$$D = -2, \quad E = -\tfrac{1}{4}, \quad \text{and} \quad F = \tfrac{5}{4}.$$

The equation of the parabola is:

$$x^2 - 2x - \tfrac{1}{4}y + \tfrac{5}{4} = 0$$

or

$$4x^2 - 8x - y + 5 = 0.$$

If $A = B = D = 0$ or $C = B = E = 0$, then we have two parallel lines, the same line twice, or no locus.

Illustration. What is the locus of the points satisfying:

$$y^2 - 5y + 6 = 0?$$

Solution. Factor the equation into:

---

* This is common practice, which is permissible since we could divide the equation by A and consider the ratios as new constants.

$$(y - 2)(y - 3) = 0$$

so that:

$$y = 2 \quad \text{or} \quad y = 3$$

for all x and we have two parallel lines.

77. The Ellipse. There are two definitions of the ellipse.

I. *An* **ellipse** *is the locus of a point which moves so that the sum of its distances from two fixed points is a constant.*

II. *The conic section for which $e < 1$ is called an* **ellipse.**

Let us refer to Fig. 45 and clearly define our terminology.

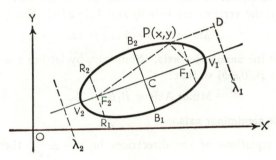

FIG. 45

F_1 and F_2 are **foci.**	C is the **center.**
V_1 and V_2 are **vertices.**	$\overline{V_1V_2}$ is the **major axis.**
λ_1 and λ_2 are **directrices.**	$\overline{B_1B_2}$ is the **minor axis.**
$PF_1/PD = e$, the **eccentricity.**	R_1R_2 is the **latus rectum.**

We note that many of the elements occur in pairs and that the ellipse is a closed curve of oval shape. The equation of the ellipse can be derived from either definition. Let us choose a special case in which we pick the two fixed points at $F_1(c,0)$ and $F_2(-c,0)$ and the general point $P(x,y)$. We see that $\overline{F_1F_2} = 2c$ and the center is at $(0,0)$. Let the constant be $2a$ so that in definition I we have $PF_1 + PF_2 = 2a$, and in terms of the coordinates:

$$\sqrt{(x - c)^2 + y^2} + \sqrt{(x + c)^2 + y^2} = 2a.$$

If we transpose one of the radicals, square both sides, simplify, and square both sides again, we have:

$$(a^2 - c^2)x^2 + a^2y^2 = a^2(a^2 - c^2).$$

To simplify further, let:

$$a^2 - c^2 = b^2.$$

Then the equation becomes:

$$b^2x^2 + a^2y^2 = a^2b^2.$$

If we divide both sides by a^2b^2, we have:

$$\frac{x^2}{a^2} + \frac{y^2}{b^2} = 1,$$

which is a *standard form* for the equation of an ellipse. If we let $y = 0$ in this equation, the x-intercepts are $x = \pm a$ so that the coordinates of the vertices are $V_1(a,0)$ and $V_2(-a,0)$. Thus:

$$\text{Major Axis} = \overline{V_1V_2} = 2a,$$

so that **a** is the **semimajor axis**. Similarly, by letting $x = 0$ we find $B_1(0,b)$ and $B_2(0,-b)$ so that:

$$\text{Minor Axis} = \overline{B_1B_2} = 2b,$$

and **b** is the **semiminor axis**.

Let the equations of the directrices be $x = \pm \dfrac{a}{e}$; then the coordinates of the point D are $\left(\dfrac{a}{e}, y\right)$ and we have:

$$\frac{PF_1}{PD} = \frac{\sqrt{(x - c)^2 + y^2}}{\sqrt{\left(x - \dfrac{a}{e}\right)^2}} = e.$$

If we now let $e = \dfrac{c}{a} < 1$ we satisfy definition II and the above equation reduces to:

$$b^2x^2 + a^2y^2 = a^2b^2$$

by squaring both sides and using $a^2 - c^2 = b^2$, so that the standard form also satisfies the general definition of a conic section given as II in Section 75. Consequently, for an ellipse with the center at the origin and foci $F_1(c,0)$, $F_2(-c,0)$ we satisfy all definitions by choosing:

$$e = \frac{c}{a} = \frac{\sqrt{a^2 - b^2}}{a},$$

and the equations of the directrices $x = \pm \dfrac{a}{e}$, and we may note $c = ae$

so that the coordinates of the foci may be written $F_1(ae,0)$ and $F_2(-ae,0)$.

The coordinates of R_1 and R_2 are obtained by letting $x = c$ in the equation of the ellipse and solving for y. Thus:

$$y = \pm \frac{b}{a}\sqrt{a^2 - c^2} = \pm \frac{b^2}{a}$$

and we have $R_1\left(c, \frac{b^2}{a}\right)$ and $R_2\left(c, -\frac{b^2}{a}\right)$. The *length of the latus rectum* then is:

$$R_1R_2 = \sqrt{\left(\frac{b^2}{a} + \frac{b^2}{a}\right)^2} = \frac{2b^2}{a}.$$

The letters a, b, and c are arbitrary constants which, however, we have chosen so that they have particular meanings for the ellipse. If we further choose a such that a^2 is the larger number of the two denominators, it will always denote the semimajor axis for the ellipse. We may now vary the standard form for the equation of an ellipse by varying the center and major axis. In summary form we have:

EQUATION	CEN-TER	VERTICES	FOCI	DIREC-TRICES
Major Axis Parallel to X-axis				
$\dfrac{x^2}{a^2} + \dfrac{y^2}{b^2} = 1$	$(0,0)$	$(a,0)$; $(-a,0)$	$(ae,0)$; $(-ae,0)$	$x = \pm \dfrac{a}{e}$
$\dfrac{(x-h)^2}{a^2} + \dfrac{(y-k)^2}{b^2} = 1$	(h,k)	$(h+a,k)$ $(h-a,k)$	$(h+ae,k)$ $(h-ae,k)$	$x = h \pm \dfrac{a}{e}$
Major Axis Parallel to Y-axis				
$\dfrac{x^2}{b^2} + \dfrac{y^2}{a^2} = 1$	$(0,0)$	$(0,a)$; $(0,-a)$	$(0,ae)$; $(0,-ae)$	$y = \pm \dfrac{a}{e}$
$\dfrac{(x-h)^2}{b^2} + \dfrac{(y-k)^2}{a^2} = 1$	(h,k)	$(h,k+a)$ $(h,k-a)$	$(h,k+ae)$ $(h,k-ae)$	$y = k \pm \dfrac{a}{e}$
Semimajor axis: a. Semiminor axis: b. Eccentricity: $e = \dfrac{\sqrt{a^2 - b^2}}{a} < 1$. Length of latus rectum: $\dfrac{2b^2}{a}$.				

The student should sketch each of the above four curves and locate the essential points.

Illustration. Find the equation of the ellipse with center at (1,3), major axis = 10 and parallel to the X-axis, and semiminor axis = 3. Find also the eccentricity, the foci, the vertices, the equations of the directrices, and the length of the latus rectum.

Solution. From the given data we have $h = 1$, $k = 3$, $a = 5$, and $b = 3$. Thus:

Equation:
$$\frac{(x-1)^2}{25} + \frac{(y-3)^2}{9} = 1.$$

Eccentricity:
$$e = \frac{\sqrt{a^2 - b^2}}{a} = \frac{\sqrt{25-9}}{5} = \frac{4}{5}.$$
$$ae = 5(\tfrac{4}{5}) = 4.$$

Foci: $F_1(5,3)$; $F_2(-3,3)$.

Vertices: $V_1(6,3)$; $V_2(-4,3)$.

Directrices:
$$x = 1 \pm \frac{5}{\frac{4}{5}} = 1 \pm \frac{25}{4},$$
$$x = \tfrac{29}{4} \quad \text{and} \quad x = -\tfrac{21}{4}.$$

Length of L.R.:
$$\frac{2b^2}{a} = \frac{2(9)}{5} = \frac{18}{5}.$$

The student should sketch the curve.

If in the general equation of the second degree:

$$Ax^2 + Bxy + Cy^2 + Dx + Ey + F = 0,$$

$B = 0$ and A and C have *like* signs, then the equation represents an ellipse.

Illustration. Reduce the equation:

$$25x^2 + 16y^2 + 100x - 32y - 284 = 0$$

to a standard form and give the elements of the curve.

Solution. Complete the square on x and y:

$$25(x^2 + 4x + 4) + 16(y^2 - 2y + 1) = 284 + 25(4) + 16(1);$$
$$25(x+2)^2 + 16(y-1)^2 = 400.$$

Divide by 400.

$$\frac{(x+2)^2}{16} + \frac{(y-1)^2}{25} = 1.$$

This is the standard form of the equation of the ellipse, which has its major axis parallel to the Y-axis. The elements are:

Semimajor axis: $a = 5$.

Semiminor axis: $b = 4$.

Center: $(-2,1)$; $h = -2$; $k = 1$.
Eccentricity: $e = \frac{3}{5}$; $ae = 3$.
Foci: $F_1(-2,4)$; $F_2(-2,-2)$.
Vertices: $V_1(-2,6)$; $V_2(-2,-4)$.
Length of L.R.: $\frac{32}{5}$.
Directrices: $y = 1 \pm \frac{25}{3}$.

The ellipse is shown in Fig. 46.

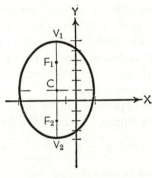

Fig. 46

78. The Hyperbola. There are two definitions of the hyperbola.

I. *A* **hyperbola** *is the locus of a point which moves so that the difference of its distances from two fixed points is a constant.*

II. *The conic section for which* $e > 1$ *is called a* **hyperbola.** Let us refer to Fig. 47 for our definitions.

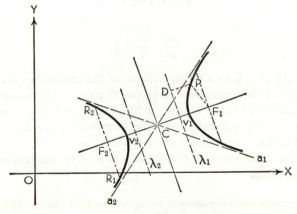

Fig. 47

F_1 and F_2 are **foci**.

V_1 and V_2 are **vertices**.

λ_1 and λ_2 are **directrices**.

a_1 and a_2 are **asymptotes**.

C is the **center**.

$\overline{V_1V_2}$ is the **transverse axis**.

$PF_1/PD = e$, the **eccentricity**.

$\overline{R_1R_2}$ is the **latus rectum**.

Let:

$$\overline{F_1F_2} = 2c \quad \text{and} \quad PF_2 - PF_1 = 2a.$$

We note that the curve is not a closed curve but has two branches. The term *asymptote* is new; it refers to the tangent through C whose point of contact is at an infinite distance from C. We note that the hyperbola gets closer and closer to its asymptotes and approaches indefinitely near to them in the neighborhood of infinity. This is the rough concept of asymptotes.*

The equation of the hyperbola can be derived in the same manner as the equation for the ellipse. From definition I we have:

$$\sqrt{(x - c)^2 + y^2} - \sqrt{(x + c)^2 + y^2} = 2a,$$

which upon simplifying becomes:

$$(a^2 - c^2)x^2 + a^2y^2 = a^2(a^2 - c^2).$$

Now since $PF_2 - PF_1 < F_1F_2$ we have $2a < 2c$ and $a^2 - c^2$ is a negative number. Thus let:

$$a^2 - c^2 = -b^2$$

and we have:

$$b^2x^2 - a^2y^2 = a^2b^2$$

or

$$\frac{x^2}{a^2} - \frac{y^2}{b^2} = 1.$$

The length $2b$ defines another axis called the **conjugate** axis. It does not intersect the curve and is perpendicular to the transverse axis. The rectangle of sides $2a$ and $2b$ locates the asymptotes as the diagonals of this rectangle. All the elements of the hyperbola are derived in the same manner as we did for the ellipse.

The standard forms of the hyperbola may be summarized in the following table.

---

* For a more thorough discussion of asymptotes see C. O. Oakley, *Analytic Geometry* (New York: Barnes & Noble, Inc., 1957), pp. 34ff.

EQUATION	CEN-TER	VERTICES	FOCI	DIREC-TRICES	ASYMPTOTES
Transverse Axis Parallel to the X-axis					
$\dfrac{x^2}{a^2} - \dfrac{y^2}{b^2} = 1$	$(0,0)$	$(a,0)$ $(-a,0)$	$(ae,0)$ $(-ae,0)$	$x = \pm\dfrac{a}{e}$	$y = \pm\dfrac{b}{a}x$
$\dfrac{(x-h)^2}{a^2} - \dfrac{(y-k)^2}{b^2} = 1$	(h,k)	$(h+a,k)$ $(h-a,k)$	$(h+ae,k)$ $(h-ae,k)$	$x = h \pm\dfrac{a}{e}$	$y - k = \pm\dfrac{b}{a}(x-h)$
Transverse Axis Parallel to the Y-axis					
$-\dfrac{x^2}{b^2} + \dfrac{y^2}{a^2} = 1$	$(0,0)$	$(0,a)$ $(0,-a)$	$(0,ae)$ $(0,-ae)$	$y = \pm\dfrac{a}{e}$	$y = \pm\dfrac{a}{b}x$
$-\dfrac{(x-h)^2}{b^2} + \dfrac{(y-k)^2}{a^2} = 1$	(h,k)	$(h,k+a)$ $(h,k-a)$	$(h,k+ae)$ $(h,k-ae)$	$y = k \pm\dfrac{a}{e}$	$y - k = \pm\dfrac{a}{b}(x-h)$

Semitransverse axis: a.

Semiconjugate axis: b.

Eccentricity: $e = \dfrac{\sqrt{a^2 + b^2}}{a} > 1$

Length of latus rectum: $\dfrac{2b^2}{a}$.

Illustration. Find the equation of the hyperbola with center at (1,3), transverse axis = 8 and parallel to the X-axis, and semiconjugate axis = 3. Find also the eccentricity, the foci, the vertices, the equation of the directrices, the equations of the asymptotes, and the length of the latus rectum.

Solution. From the given data we have $h = 1$, $k = 3$, $a = 4$, and $b = 3$. Thus:

Equation:　　　　$\dfrac{(x-1)^2}{16} - \dfrac{(y-3)^2}{9} = 1.$

Eccentricity:　　　　$e = \dfrac{\sqrt{a^2 + b^2}}{a} = \dfrac{\sqrt{25}}{4} = \dfrac{5}{4}.$

　　　　　　　　$ae = 5.$

Foci:　　　　$F_1(6,3)$; $F_2(-4,3).$

Vertices:　　　　$V_1(5,3)$; $V_2(-3,3).$

Directrices:　　　　$x = 1 \pm \frac{16}{5}.$

Asymptotes:　　　　$y - 3 = \pm\frac{3}{4}(x - 1)$　or

　　　　　　　　$4y - 12 = \pm(3x - 3).$

Length of L.R.:　　　　$\frac{18}{4} = \frac{9}{2}.$

See Fig. 48 for sketch of curve.

If in the general equation of the second degree:

$$Ax^2 + Bxy + Cy^2 + Dx + Ey + F = 0,$$

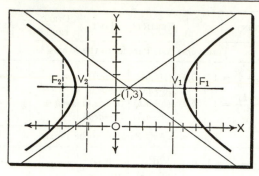

Fig. 48

$B = 0$ and A and C have *opposite* signs, then the equation represents a hyperbola.

Illustration. Reduce the equation:

$$-16x^2 + 9y^2 + 48x - 36y = 144$$

to a standard form and give the elements of the curve.

Solution. Complete the square on x and y:

$$-16(x^2 - 3x + \tfrac{9}{4}) + 9(y^2 - 4y + 4) = 144 - \tfrac{9}{4}(16) + 9(4);$$
$$-16(x - \tfrac{3}{2})^2 + 9(y - 2)^2 = 144.$$

Divide by 144:

$$-\frac{(x - \tfrac{3}{2})^2}{9} + \frac{(y - 2)^2}{16} = 1.$$

This is the standard form of a hyperbola which has its transverse axis parallel to the Y-axis. The elements of the curve are:

Center: $(\tfrac{3}{2},2)$; $h = \tfrac{3}{2}$; $k = 2$.
Semitransverse axis: $a = 4$.
Semiconjugate axis: $b = 3$.
Eccentricity: $e = \tfrac{5}{4}$; $ae = 5$.
Foci: $F_1(\tfrac{3}{2},7)$; $F_2(\tfrac{3}{2},-3)$.
Vertices: $V_1(\tfrac{3}{2},6)$; $V_2(\tfrac{3}{2},-2)$.
Directrices: $y = 2 \pm \tfrac{16}{5}$.
Asymptotes: $y - 2 = \pm\tfrac{4}{3}(x - \tfrac{3}{2})$ or
$y - 2 = \pm(\tfrac{4}{3}x - 2)$.
Length of L.R.: $\tfrac{18}{4} = \tfrac{9}{2}$.

The student should sketch the curve.

79. Conjugate and Rectangular Hyperbolas. In discussing the hyperbola we defined the transverse and conjugate axes. If we arbitrarily interchange the role of these axes we are led to another hyperbola and speak of conjugate hyperbolas. **Conjugate hyperbolas** are concentric hyperbolas such that the transverse axis of one coincides with the conjugate axis of the other. The standard forms of the equations are obtained by changing the signs in the variable terms. Thus, the following are conjugate hyperbolas:

$$\begin{cases} \dfrac{(x-h)^2}{a^2} - \dfrac{(y-k)^2}{b^2} = 1, \\[2ex] -\dfrac{(x-h)^2}{a^2} + \dfrac{(y-k)^2}{b^2} = 1. \end{cases}$$

It is easily seen that conjugate hyperbolas have the same asymptotes.

Illustration. Sketch the hyperbola $4x^2 - y^2 - 36 = 0$. Find the equation of the conjugate hyperbola and sketch it on the same figure.

Solution. The standard form of the given hyperbola is:

$$\frac{x^2}{9} - \frac{y^2}{36} = 1.$$

The elements are:

$a = 3$; $b = 6$; $e = \sqrt{5}$; $ae = 3\sqrt{5}$; $V_1(3,0)$; $V_2(-3,0)$; $F_1 = (3\sqrt{5},0)$; $F_2(-3\sqrt{5},0)$; asymptotes: $y = \pm 2x$; directrices: $x = \pm\frac{3}{5}\sqrt{5}$; length of L.R. = 24. The conjugate hyperbola is:

$$-\frac{x^2}{9} + \frac{y^2}{36} = 1.$$

The elements are:

$a = 6$; $b = 3$; $e = \frac{1}{2}\sqrt{5}$; $ae = 3\sqrt{5}$; $V'_1(0,6)$; $V'_2(0,-6)$; $F'_1(0,3\sqrt{5})$; $F'_2(0,-3\sqrt{5})$; asymptotes: $y = \pm 2x$; directrices: $y = \pm\frac{12}{5}\sqrt{5}$; length of L.R. = 3. The sketch is shown in Fig. 49, p. 130.

If $a = b$, the hyperbola is called a **rectangular hyperbola** or an **equilateral hyperbola.** In this case the asymptotes are perpendicular to each other. This leads to a special form of the equation of a hyperbola, namely, one in which the asymptotes are the coordinate axes. The form is:

$$xy = k,$$

where k is a constant. By giving values to one of the variables we can find the values of the other easily. Plotting such points will result in a hyperbola.

Illustration. Draw the graph of $xy = 6$.
Solution. Calculate the table of values.

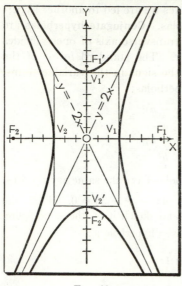

x	± 1	± 2	± 3	± 4	± 6
y	± 6	± 3	± 2	$\pm \frac{3}{2}$	± 1

Fig. 50 shows the graph.

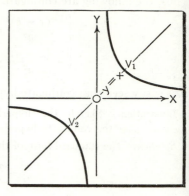

FIG. 49   FIG. 50

80. Transformations of Coordinates. We have seen that the general quadratic equation of the second degree:

$$(1) \qquad Ax^2 + Bxy + Cy^2 + Dx + Ey + F = 0$$

represents a conic if $B = 0$. It may be suspected that it always represents a conic even if $B \neq 0$, and this is true. We proceed to establish this fact by showing that we can remove the xy term by a shift of the coordinate axes, and in so doing we shall discuss transformations of coordinates in general. A **transformation** is the substitution of new variables for those occurring in the given problem. The substitutions are made to simplify the equations or to display prominently a particular property. One of the simplest is known as a **translation** of the axes parallel to themselves. It is obtained by the following *equations of transformation:*

$$(2) \qquad \begin{cases} x = x' + h, \\ y = y' + k, \end{cases}$$

where (h,k) are the coordinates, with respect to the old axes, of the origin for the new axes. See Fig. 51.

A translation may be used to remove the first-degree terms from the quadratic equation.

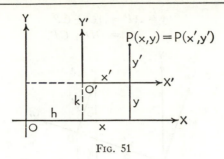

Fig. 51

Illustration. Remove the first-degree terms from:

$$x^2 + y^2 - 6x + 4y = 3.$$

Solution. Substitute $x = x' + h$ and $y = y' + k$ into the equation and determine h and k so that the new equation has no first-degree terms. Thus:

$$(x' + h)^2 + (y' + k)^2 - 6(x' + h) + 4(y' + k) = 3$$

or

$$x'^2 + 2hx' + h^2 + y'^2 + 2ky' + k^2 - 6x' - 6h + 4y' + 4k = 3.$$

Now if there are to be no first-degree terms, the coefficients of x' and y' must be zero; i.e.:

$$2h - 6 = 0 \quad \text{or} \quad h = 3,$$
$$2k + 4 = 0 \quad \text{or} \quad k = -2.$$

The equation then becomes;

$$x'^2 + 6x' + 9 + y'^2 - 4y' + 4 - 6x' - 18 + 4y' - 8 = 3$$

or

$$x'^2 + y'^2 = 16,$$

which is a circle of radius 4 and has a center at $(3, -2)$ referred to the original axes.

The second transformation that we shall consider is the one which **rotates** the axes counterclockwise about the origin. The *equations of transformation* are:

(3)
$$\begin{cases} x = x' \cos \phi - y' \sin \phi, \\ y = x' \sin \phi + y' \cos \phi. \end{cases}$$

These equations are obtained by considering Fig. 52 and noting:

$$x = OA = OB - AB$$
$$= OB - CD;$$

$$y = AP = AC + CP$$
$$= BD + CP.$$

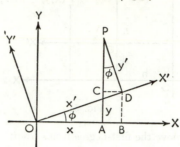

Fig. 52

Now, from Section 22:

$$OB = x' \cos \phi; \quad CD = y' \sin \phi;$$
$$BD = x' \sin \phi; \quad CP = y' \cos \phi;$$

and the transformation equations (3) are obtained by a substitution.

*A rotation is used to remove the **xy** term from an equation of the second degree.*

Let us substitute the rotation equations (3) into (1) and focus our attention on the coefficient of the xy term.

From the x^2 term we get $(-2A \sin \phi \cos \phi)x'y'$;

From the xy term we get $B(\cos^2 \phi - \sin^2 \phi)x'y'$;

From the y^2 term we get $(2C \sin \phi \cos \phi)x'y'$, so that the complete coefficient of $x'y'$ is:

$$-(A - C)2 \sin \phi \cos \phi + B(\cos^2 \phi - \sin^2 \phi).$$

It is desired to make this coefficient zero. Making use of the formulas (see Section 34):

$$2 \sin \phi \cos \phi = \sin 2\phi \quad \text{and} \quad \cos^2 \phi - \sin^2 \phi = \cos 2\phi$$

we have:

(4)
$$\frac{\sin 2\phi}{\cos 2\phi} = \tan 2\phi = \frac{B}{A - C}.$$

Equation (4) defines the angle which will rotate the axes to eliminate the xy term. The trigonometric functions needed for the transformation equations (3) are found from:

$$(5) \quad \begin{cases} \sin\ \phi = \sqrt{\dfrac{1 - \cos 2\phi}{2}}, \\[3mm] \cos\ \phi = \sqrt{\dfrac{1 + \cos 2\phi}{2}}, \end{cases}$$

and we limit ourselves to angles less than 90°.

Illustration. Remove the xy term and sketch the curve:

$$16x^2 + 24xy + 9y^2 + 15x - 20y = 75.$$

Solution. First determine the angle ϕ for the rotation.

$$\tan 2\phi = \frac{B}{A - C} = \frac{24}{7},$$
$$\cos 2\phi = \tfrac{7}{25},$$
$$\sin \phi = \sqrt{\tfrac{1}{2}(1 - \tfrac{7}{25})} = \tfrac{3}{5},$$
$$\cos \phi = \sqrt{\tfrac{1}{2}(1 + \tfrac{7}{25})} = \tfrac{4}{5}.$$

The transformation equations are:

$$x = \tfrac{1}{5}(4x' - 3y'),$$
$$y = \tfrac{1}{5}(3x' + 4y'),$$

and the given equation is reduced to:

$$x'^2 - y' = 3$$

or in standard form:

$$x'^2 = y' + 3,$$

which is a parabola. The sketch is shown in Fig. 54.

Fig. 53

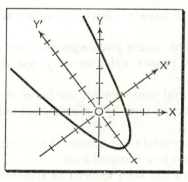

Fig. 54

If in the original equation $A = C$, rotate the axes through $\phi = 45°$.
This is easily seen, as then:

$$\tan 2\phi = \frac{B}{A - C} = \frac{B}{0} = \infty$$

and $2\phi = 90°$ so that $\phi = 45°$.

81. Summary. It should be clear by now that the general equation of the second degree in two variables:

(1) $$Ax^2 + Bxy + Cy^2 + Dx + Ey + F = 0$$

represents a conic. Let us define the **discriminant** of equation (1) to be:

$$\Delta = \begin{vmatrix} 2A & B & D \\ B & 2C & E \\ D & E & 2F \end{vmatrix}$$

If $\Delta = 0$, the conic is degenerate and the locus may be parallel lines, intersecting lines, or a point.

If $\Delta \neq 0$, we have a proper conic. The proper conics may be classified by referring to the following table:

Proper Conics

Condition	Locus
$B^2 - 4AC = 0$	A parabola
$B^2 - 4AC < 0$	An ellipse
$B^2 - 4AC > 0$	A hyperbola

The conics which have a center are called **central** conics and $B^2 - 4AC \neq 0$.

To obtain the locus from a given equation, first, test the discriminant to see if it is a proper conic; secondly, test $B^2 - 4AC$ to see if it is a central conic.

(a) If it is a central conic remove the linear terms by a translation, then remove the xy term by a rotation and reduce the equation to a standard form.

(b) If it is not a central conic remove the xy term and reduce the transformed equation to a standard form.

Once the equation has been reduced to standard form, calculate the elements and sketch the curve.

Illustration. Construct the locus of:

$$5x^2 + 6xy + 5y^2 - 38x - 42y + 69 = 0. \tag{1}$$

Solution.

$$\Delta = \begin{vmatrix} 10 & 6 & -38 \\ 6 & 10 & -42 \\ -38 & -42 & 138 \end{vmatrix} = -4096 \neq 0.$$

∴   Conic is proper.

$$B^2 - 4AC = 36 - 100 = -64 < 0.$$

∴   Conic is an ellipse, a central conic.

Remove linear terms by substituting:

$$x = x' + h \quad \text{and} \quad y = y' + k$$

and set coefficients of x' and y' equal to zero.

Coefficient of x': $10h + 6k - 38 = 0.$
Coefficient of y': $6h + 10k - 42 = 0.$
Thus: $h = 2 \quad \text{and} \quad k = 3.$

Substitute $x = x' + 2$ and $y = y' + 3$ into (1):

$$5x'^2 + 6x'y' + 5y'^2 - 32 = 0. \tag{2}$$

Remove the $x'y'$ term by a rotation:

$$\tan 2\phi = \frac{6}{5 - 5} = \infty.$$

∴ $\phi = 45°$, $\sin \phi = \dfrac{1}{\sqrt{2}}$,   and $\cos \phi = \dfrac{1}{\sqrt{2}}.$

The transformation equations are:

$$x' = \frac{1}{\sqrt{2}}(x'' - y''), \quad y' = \frac{1}{\sqrt{2}}(x'' + y'').$$

Substitute these into (2) and simplify:

$$4x''^2 + y''^2 = 16. \tag{3}$$

A division by 16 puts (3) into standard form:

$$\frac{x''^2}{4} + \frac{y''^2}{16} = 1.$$

We have now translated the origin to (2,3) and rotated through 45° to new axes CX'' and CY''. Referring to *these axes* the conic is an ellipse with major axes parallel to CY''.

$a = 4, b = 2, ae = 2\sqrt{3}, V_1(0,4), V_2(0,-4), F_1(0,2\sqrt{3}), F_2(0,-2\sqrt{3}).$

Length of L.R. = 2.   The conic is shown in Fig. 55, p. 136.

82. Exercise VII.

1. Calculate the semiaxes, eccentricity, vertices, foci, length of latus

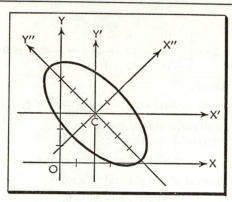

rectum, and plot the curve:

(a) $x^2 + 25y^2 = 25$.
(b) $2y^2 = 5x$.
(c) $9x^2 - 16y^2 = 144$.
(d) $x^2 + y^2 - 25 = 0$.
(e) $y^2 - 2y + 1 = 4x - 12$.

2. Transform into standard form and plot the curve:

(a) $4x^2 + 4y^2 - 12x - 20y = 30$.
(b) $3x^2 + 4y^2 - 12 = 0$.
(c) $y^2 + 2\sqrt{3}xy - x^2 = 4$.
(d) $4x^2 + y^2 + 24x - 16y + 84 = 0$.
(e) $4x^2 + 4xy + y^2 - 6x + 12y = 0$.

3. Find the vertex and focus and plot the curve:

$$y = 6x^2 - 11x + 7.$$

4. Find the equation of the ellipse whose minor axis is 48 and whose foci are at the points in which the circle $x^2 + y^2 = 49$ meets the X-axis. Sketch the ellipse.

5. Find the equation of the hyperbola with center at the origin, which passes through the point $(-2,3)$, and whose asymptotes are given by $y = \pm 2x$. Find the conjugate of this hyperbola.

8

POLAR COORDINATES

83. Definitions. We shall now consider another method of representing points in a plane by coordinates. Let there be given a fixed line OX, usually horizontal, and a point O on this line. The line OX is called the **polar axis,** and the point O is called the **pole.** The position of any point P is determined by the length $OP = r$ (also represented by ρ) and the angle $XOP = \theta$. The length r is called the **radius vector;** the angle θ is called the **polar angle;** and together they are called the **polar coordinates.** We write $P(r,\theta)$. The polar angle is *positive* when measured *counterclockwise.* The radius vector is *positive* if it is measured on the *terminal side of θ.* A *negative* radius vector is measured on the terminal side of θ *produced through O.* Fig. 56 shows three points.

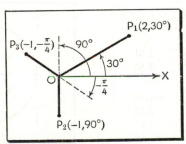

Fig. 56

Since an angle in a plane can have more than one designation ($\theta = 30° = 390° = -330° = \cdots$) every point has an infinite number of polar coordinates. To be more definite we shall restrict ourselves to $0 \leq \theta \leq 360°$ unless specifically stated otherwise. To ease the plotting of points special graph paper, called *polar coordinate paper,* is available. Fig. 57 shows some points plotted on this graph paper. The concentric circles about the origin measure units of r, and the angles are indicated.

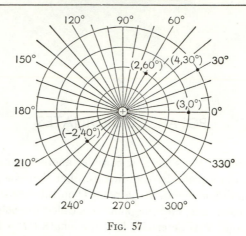

Fig. 57

If r and θ are connected by an equation, values may be assigned to θ and corresponding values for r may be calculated. We then have a table of values for points which may be plotted and joined by a curve, thus describing the locus of the equation.

Illustration. Draw the graph of $r = 3 \cos \theta$.
Solution. Assign values to θ and calculate the corresponding values of r:

θ	$\cos \theta$	$r = 3 \cos \theta$
0	1.00	3
30°	.87	2.61
60°	.50	1.50
90°	0	0
120°	−.50	−1.50
150°	−.87	−2.61
180°	−1.00	−3

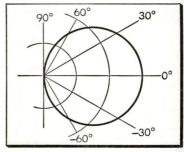

Fig. 58

The points are plotted and joined by a smooth curve to get the graph of Fig. 58.

If an equation contains only *one* of the variables (r,θ), it means that the other variable may have any and all values.

Illustration. What is the graph of $r = 5$?
Solution. This equation says that for *all values of* θ, $r = 5$; i.e., $(5,0°)$, $(5,30°)$, etc. are all points on the graph, which then is a circle of radius 5 with the center at the origin.

In plotting equations in polar coordinates, it is convenient to be aware of the *symmetry* of the curve. Thus, if we can replace θ by $-\theta$ and obtain the same value of r, then the locus is said to be *symmetric with respect to the polar axis*. If we can replace r by $-r$ for the same value of θ, then the curve is said to be *symmetric with respect to the pole*.

Illustration. Does the curve $r^2 = a^2 \cos \theta$ have any symmetry?
Solution. Since $\cos \theta = \cos(-\theta)$, the curve is symmetric with respect to the polar axis. Furthermore, since we have values $+r$ and $-r$ for each θ the curve is symmetric with respect to the pole.

84. Single-valued Functions. The simplest problem of tracing polar curves is the case in which there is only one value of r for each value of θ. These curves are called *single-valued functions*.

Illustration. Trace the curve $r = 5 \cos 2\theta$.

Solution. Calculate a table of values.

θ	0°	15°	22½°	30°	45°	60°
2θ	0°	30°	45°	60°	90°	120°
r	5	$\frac{5}{2}\sqrt{3}$	$\frac{5}{2}\sqrt{2}$	$\frac{5}{2}$	0	$-\frac{5}{2}$

It is easily seen that as θ varies from 45° to 90°, 2θ varies from 90° to 180°, and we obtain the negative values of r in reverse order from the above table. Plotting the points we obtain the half-loops marked 1 and 2 on Fig. 59. We also note that the curve is symmetric with respect to the polar axis. Continuing for all values of θ up to 360° and using the symmetric property, we obtain the entire curve of Fig. 59. Note the order of the half-loops.

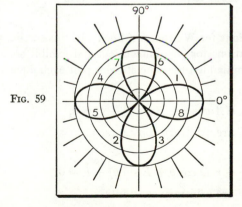

Fig. 59

85. Double-valued Functions. The case in which r^2 is expressed as a function of θ yields two values of r for each value of θ and is called a *double-valued function*.

Illustration. Trace the curve $r^2 = 25 \sin 2\theta$.

Solution. Calculate a table of values.

θ	$0°$	$15°$	$22\frac{1}{2}°$	$30°$	$45°$	$60°$
2θ	$0°$	$30°$	$45°$	$60°$	$90°$	$120°$
r	0	$\pm\frac{5}{2}\sqrt{2}$	$\pm\dfrac{5}{\sqrt[4]{2}}$	$\pm 5\dfrac{\sqrt[4]{3}}{\sqrt{2}}$	± 5	$\pm 5\dfrac{\sqrt[4]{3}}{\sqrt{2}}$
r	0	± 3.54	± 4.20	± 4.65	± 5	± 4.65

As we continue the values of θ we see that the numerical values of r are continued and, of course, the curve is symmetric with respect to the pole. The complete plot is shown in Fig. 60.

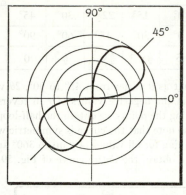

Fᴉɢ. 60

86. Polar Equations for Well-known Loci. In this section we shall list the polar equations for well-known loci and exhibit some of the curves. It is recommended that the student plot each equation.

I. *Straight Line.*

$$r(A \cos \theta + B \sin \theta) + C = 0,$$

where A, B, and C are arbitrary constants.

II. *Circle.*

$$r^2 + r(D \cos \theta + E \sin \theta) + F = 0,$$

where D, E, and F are arbitrary constants.

III. *Conics.*

$$r = \frac{ed}{1 \pm e \cos \theta},$$

where e is the eccentricity, d is the distance from the focus to the directrix, the focus is at the pole, and the polar axis is perpendicular to the directrix.

IV. *Three-leaved Rose.*

$r = a \cos 3\theta.$  $r = a \sin 3\theta.$

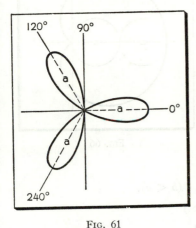

Fig. 61

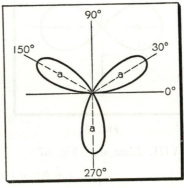

Fig. 62

V. *Four-leaved Rose.*

$r = a \cos 2\theta.$ $r = a \sin 2\theta.$

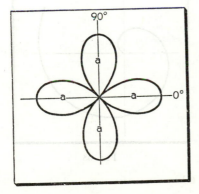

Fig. 63

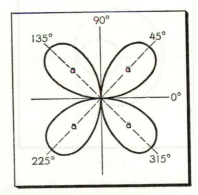

Fig. 64

VI. *Lemniscate of Bernoulli.* (Fig. 65.)

$$r^2 = a^2 \cos 2\theta.$$

(See also Fig. 60.)

VII. *Cardioid.* (Fig. 66.)

$$r = a(1 - \cos \theta).$$

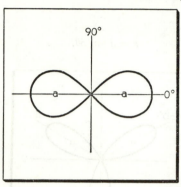

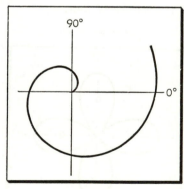

FIG. 65

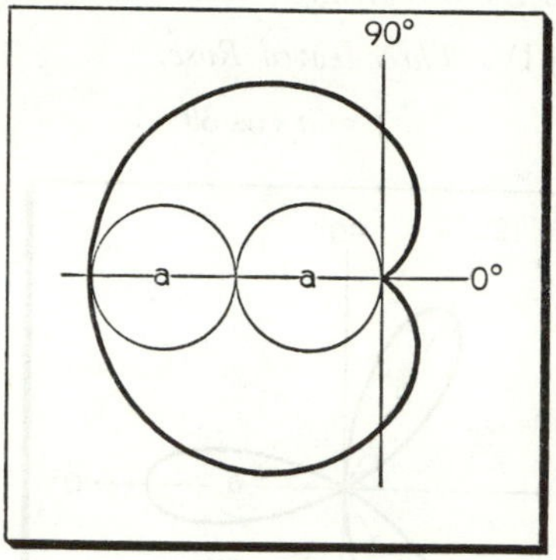

FIG. 66

VIII. *Limaçon.* (Fig. 67.)

$$r = b - a \cos \theta, \quad (b < a).$$

IX. *Spiral of Archimedes.* (Fig. 68.)

$$r = a\theta.$$

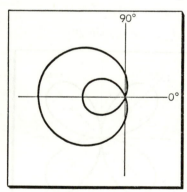

FIG. 67

FIG. 68

X. *Conchoid of Nicomedes.* (Fig. 69.)

$$r = a \csc \theta + b.$$

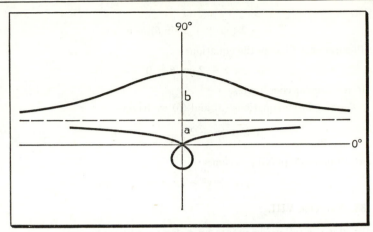

FIG. 69

87. Transformation of Axes. It is often desired to transform the equation of a curve from its given form in rectangular coordinates to a form in polar coordinates or vice versa. Let the pole be at the origin and let the polar axis coincide with the positive X-axis (see Fig. 70). Then we have:

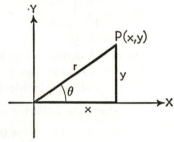

FIG. 70

(1) $x = r \cos \theta$.

(2) $y = r \sin \theta$.

(3) $r^2 = x^2 + y^2$.

(4) $\tan \theta = \dfrac{y}{x}$.

(5) $\sin \theta = \dfrac{y}{\pm \sqrt{x^2 + y^2}}$.

(6) $\cos \theta = \dfrac{x}{\pm \sqrt{x^2 + y^2}}$.

Illustration. Change the equation:

$$x^2 + y^2 - 2x - 6y = 6$$

to polar coordinates.

Solution. From equations (1), (2), and (3) we have:

$$x^2 + y^2 = r^2,$$
$$-2x = -2r \cos \theta,$$
$$-6y = -6r \sin \theta.$$

Thus the polar equation is:

$$r^2 - 2r \cos \theta - 6r \sin \theta = 6,$$

or

$$r^2 - 2r(\cos \theta + 3 \sin \theta) - 6 = 0.$$

Illustration. Change the equation:

$$r + 2 \cos \theta = 0$$

to rectangular coordinates.

Solution. From equations (3) and (6) we have:

$$\sqrt{x^2 + y^2} + 2 \frac{x}{\sqrt{x^2 + y^2}} = 0,$$

which upon simplifying becomes:

$$x^2 + y^2 + 2x = 0 \qquad \text{(circle)}.$$

88. Exercise VIII.

1. Plot the following points on polar coordinate paper.

$(3,30°)$; $\left(-3, \dfrac{\pi}{2}\right)$; $\left(2, \dfrac{\pi}{6}\right)$; $(-1,-45°)$;

$(\tfrac{3}{2},120°)$; $(-1,-180°)$; $(6,330°)$.

2. Sketch the following curves:

(a) $r = \dfrac{1}{1 - \cos \theta}$.

(b) $r = \sin \tfrac{1}{2}\theta$.

(c) $r\theta = 5$ (hyperbolic spiral).

(d) $r = a(4 \cos \theta - \sec \theta)$ (trisectrix).

(e) $r = a \sin 4\theta$ (eight-leaved rose).

3. Transform the following equations into polar coordinates:

(a) $x^2 + y^2 + 4x - 8y = 0$.

(b) $4x^2 - y^2 = 16$.

(c) $y^2 = x^3$.

4. Transform the following equations into rectangular equations:

(a) $r(3 \cos \theta - 5 \sin \theta) + 7 = 0$.

(b) $r = \dfrac{5}{2 - 2 \cos \theta}$.

(c) $r^2 = 9 \cot \theta$.

5. Find the polar equation of the path of a point which moves so that its distance from the pole is equal to its distance from the line perpendicular to the OX axis at $(6,0)$. Sketch the curve.

9

PROGRESSIONS

89. Sequence of Numbers. It is frequently desired to order a group of objects in such a way that there is a first object, a second object, etc. For n objects we may represent the first object by a_1, the second by a_2, etc., so that we have the ordered set:

$$(1) \qquad a_1, a_2, a_3, \ldots, a_n.$$

An ordered set of objects is called a **sequence,** and the individual objects are called **terms** or **elements** of the sequence. A term, a_i, where i* may equal 1, 2, 3, or any number up to n, is called a **general term.**

These concepts may be applied to numbers. Thus we say that a *sequence of numbers* is a set of numbers arranged in a definite order. A sequence may be represented in many ways such as (1) above. Another representation is:

$$(2) \qquad a_i, \quad (i = 1,2,3,\cdots,n),$$

which is read "a_i for i equal 1 to n." We may thus condense the writing of a sequence by giving the formula for the general term as a function of i, provided, of course, that the formula exists.

Illustration 1. Write the first, second, and seventh term of the sequence:

$$a_i = 2i + 3, \quad (i = 1,2,3,\cdots,7).$$

Solution.

$$a_1 = 2(1) + 3 = 5,$$
$$a_2 = 2(2) + 3 = 7,$$
$$a_7 = 2(7) + 3 = 17.$$

Another required concept is that of finding the sum of a given number of terms of a sequence. This may be indicated by two symbols:

* In this case "i" does not represent the imaginary number.

145

$$S_n = a_1 + a_2 + a_3 + \cdots + a_n$$

and

$$\sum_{i=1}^{n} a_i = a_1 + a_2 + a_3 + \cdots + a_n.$$

The use of capital sigma, Σ, is frequent in higher mathematics, and it is a most convenient symbol. $\sum_{i=1}^{n} a_i$ is read "the summation of a_i from $i = 1$ to n."

Illustration 2. Find the sum of the first seven terms of $a_i = 2i + 3$, $(i = 1,2,\cdots,7)$.

Solution.

$$\sum_{i=1}^{7} (2i + 3) = 5 + 7 + 9 + 11 + 13 + 15 + 17 = 77.$$

90. Arithmetic Progression, A.P.

Definition. *An* **arithmetic progression** *is a sequence of numbers in which each term, after the first, is obtained from the preceding one by adding to that term a fixed number called the* **common difference.**

Let:

- a denote the first term,
- d denote the common difference,
- n denote the number of terms,
- l denote the nth (last) term,
- s denote the sum of the first n terms.

Then the A.P. is: $a, a + d, a + 2d, \ldots, l.$

The nth term is given by:

$$l = a + (n - 1)d.$$

The sum of the first n terms is given by:

$$\sum_{i=1}^{n} a_i = s = \frac{n}{2}(a + l)$$
$$= \frac{n}{2}[2a + (n - 1)d].$$

Given any two terms of an A.P.; the terms between them are called the **arithmetic means.**

If any three of the five elements a, d, n, l, and s are given, the other two may be found.

Illustration 1. Given $a = 12$, $d = 7$, and $l = 75$. Find n and s.

Solution.

$l = a + (n - 1)d.$	$s = \dfrac{n}{2}(a + l)$
$75 = 12 + (n - 1)7.$	$= \dfrac{10}{2}(12 + 75)$
$n - 1 = 9.$	$= 5(87)$
$n = 10.$	$= 435.$

Illustration 2. Given $a = 3$, $l = -47$, and $s = -242$. Find n and d.

Solution.

$s = \dfrac{n}{2}(a + l).$	$l = a + (n - 1)d.$
$2(-242) = n(3 - 47).$	$-47 = 3 + 10d.$
$n = 11.$	$d = -5.$

Illustration 3. Insert seven arithmetic means between -1 and 5.

Solution. We have $a = -1$ and $l = 5$. If we insert seven numbers between two given numbers we have $n = 7 + 2 = 9$ numbers. Now:

$$l = a + (n - 1)d.$$
$$5 = -1 + (9 - 1)d.$$
$$d = \tfrac{3}{4}.$$

Therefore the A.P. is:

$$-1, -\tfrac{1}{4}, \tfrac{1}{2}, \tfrac{5}{4}, 2, \tfrac{11}{4}, \tfrac{7}{2}, \tfrac{17}{4}, 5.$$

91. Geometric Progression, G.P.

Definition. *A **geometric progression** is a sequence of numbers in which each term, after the first, is obtained from the preceding one by multiplying that term by a fixed number called the **common ratio**.*

Let:

a denote the first term,

r denote the common ratio,

n denote the number of terms,

l denote the nth (last) term,

s denote the sum of the first n terms.

Then the G.P. is:

$$a, ar, ar^2, ar^3, \ldots, l.$$

The nth term is given by:

$$l = ar^{n-1}.$$

The sum of the first n terms is given by:

$$\sum_{i=1}^{n} a_i = s = a \frac{1 - r^n}{1 - r}$$

$$= \frac{a - rl}{1 - r}, \quad r \neq 1.$$

Given any two terms of a G.P.; the terms between them are called the **geometric means.**

If any three of the five elements a, r, l, n, and s are given, the other two may be found.

Illustration 1. Given $a = 2$, $r = 3$, and $n = 7$. Find l and s.
Solution.

$$l = ar^{n-1}$$
$$= 2(3)^{7-1}$$
$$= 1458.$$

$$s = \frac{a - rl}{1 - r}$$
$$= \frac{2 - 3(1458)}{1 - 3}$$
$$= 2186.$$

Illustration 2. Given $a = -2$, $n = 7$, and $l = -\frac{1}{32}$. Find r and s.
Solution.

$$l = ar^{n-1}.$$
$$-\frac{1}{32} = (-2)r^6.$$
$$\frac{1}{64} = r^6.$$
$$\pm\frac{1}{2} = r.$$

$$s = \frac{a - rl}{1 - r}$$
$$= \frac{-2 - (\pm\frac{1}{2})(-\frac{1}{32})}{1 - (\pm\frac{1}{2})}$$
$$= -\frac{127}{32}, \ -\frac{43}{32}.$$

Illustration 3. Insert five geometric means between 7 and 448.
Solution. $a = 7$, $l = 448$, $n = 2 + 5 = 7$.

$$l = ar^{n-1}.$$
$$448 = 7(r)^6.$$
$$64 = r^6.$$
$$\pm 2 = r.$$

There are two G.P.'s:

(a) 7, 14, 28, 56, 112, 224, 448.
(b) 7, -14, 28, -56, 112, -224, 448.

92. Infinite Sequence. So far we have been discussing sequences of numbers which have a last term. It is also possible to have sequences of numbers which do not have a last term but continue indefinitely. If after each term of a sequence there exists another term, it is called an **infinite sequence.** Consider the sequence:

$$a_i = \frac{1}{2^i}, \quad (i = 1,2,3,\cdots)$$

which can be written:

$$\tfrac{1}{2}, \tfrac{1}{4}, \tfrac{1}{8}, \tfrac{1}{16}, \tfrac{1}{32}, \cdots$$

We notice that the numbers are getting smaller but are always positive. In other words, they are getting closer and closer to zero. In such cases we say that the sequence approaches a limit.

Definition. *A sequence a_i, $(i = 1,2,\cdots,n,\cdots)$, is said to approach the constant c as a* **limit,** *if the value of $|c - a_i|$ becomes and remains less than any preassigned positive number, however small.*

If a sequence approaches a limit, it is said to be **convergent;** if it does *not* approach a limit, it is said to be **divergent.**

We are especially interested in an infinite geometric progression in which $|r| < 1$. Consider the sum:

$$a + ar + ar^2 + ar^3 + \cdots + ar^n + \cdots$$

with $|r| < 1$. The sum of the first n terms is given by:

$$S_n = a\frac{1 - r^n}{1 - r} = \frac{a}{1 - r} - \frac{ar^n}{1 - r}.$$

Now, if we continue the sequence indefinitely, the term:

$$\frac{ar^n}{1 - r}$$

becomes very small since $|r| < 1$, and approaches zero as a limit. Thus we have the sum of the infinite geometric progression approaching the first term of the above equation, and we say:

$$S = \frac{a}{1 - r}$$

for this progression.

Illustration 1. Find the sum of:

$$\tfrac{1}{2}, \tfrac{1}{4}, \tfrac{1}{8}, \tfrac{1}{16}, \tfrac{1}{32}, \cdots$$

Solution. $a = \tfrac{1}{2}$ and $r = \tfrac{1}{2}$.

$$S = \frac{a}{1 - r} = \frac{\tfrac{1}{2}}{1 - \tfrac{1}{2}} = 1.$$

Illustration 2. Find the sum of:

$$\tfrac{4}{3} + \tfrac{4}{9} + \tfrac{4}{27} + \cdots$$

Solution. $a = \tfrac{4}{3}$ and $r = \tfrac{1}{3}$.

$$S = \frac{a}{1 - r} = \frac{\frac{4}{3}}{1 - \frac{1}{3}} = 2.$$

A repeating decimal may be changed to an equivalent common fraction.

Illustration. Change the repeating decimal $0.5363636 \cdots$ into an equivalent common fraction.

Solution. Write the decimal as the sum of an infinite geometric sequence:

$$0.5363636 \cdots = .5 + .036 + .00036 + \cdots = .5 + a + \sum_{i=1}^{\infty} ar^i.$$

Thus: $a = .036, r = .01 < 1.$

$$\therefore \quad S = \frac{a}{1 - r} = \frac{.036}{1 - .01} = \frac{.036}{.99} = \frac{4}{110}.$$

The decimal is then converted to:

$$0.5363636 \cdots = \tfrac{5}{10} + \tfrac{4}{110} = \tfrac{59}{110}.$$

93. Harmonic Progression, H.P.

Definition. *A* **harmonic progression** *is a sequence of numbers whose reciprocals form an arithmetic progression.*

To find any of the elements of a harmonic progression we simply take the reciprocals of the given numbers, find the elements of the corresponding arithmetic progression, and take the reciprocals again.

Illustration 1. Find the seventh term of the H.P. $1, \frac{1}{2}, \frac{1}{3}, \ldots$.
Solution. The A.P. is:

$$1, 2, 3, \ldots$$

with $a = 1, d = 1,$ and $n = 7.$

$$l = a + (n - 1)d = 1 + 6 = 7.$$

$\therefore$ the seventh term of H.P. is $\frac{1}{7}$.

Illustration 2. Insert three harmonic means between 1 and $\frac{1}{29}$.
Solution. Insert three arithmetic means between 1 and 29; i.e., $a = 1$, $l = 29$, and $n = 5.$

$$l = a + (n - 1)d.$$
$$29 = 1 + 4d.$$
$$7 = d.$$
$$\therefore \quad \text{A.P.} \equiv 1, 8, 15, 22, 29.$$
$$\text{H.P.} \equiv 1, \tfrac{1}{8}, \tfrac{1}{15}, \tfrac{1}{22}, \tfrac{1}{29}.$$

94. Permutations. Let us consider a set of objects which are distinguishable from each other. We have already seen that an important phase of mathematics is the ordering of the elements of a set. In Section 89 we ordered the objects by assigning a first term, a second term, etc. There are other methods of ordering. The first to be considered is a permutation.

Definition. *Each different* **arrangement, a selection in a definite order,** *which can be made from a given number of objects, by taking any part or all of them at a time, is called a* **permutation.**

Fundamental Principle. If one thing can be done in p different ways, and after it has been done in any one of these ways, a second thing can be done in q different ways, then the two things can be done together, in the order stated, in pq different ways.

This principle can be generalized to a third thing in r different ways, a fourth thing in s different ways, etc., and we state that the number of different ways in which all things can be done in the order stated is:

$$p \cdot q \cdot r \cdot s \cdots \cdots$$

The symbol $P(n,r)$ is used to denote the *number of permutations of n things taken r at a time.*

We have the following formulas:

I. The permutation of n *different* things taken r at a time is:

$$P(n,r) = n(n-1)(n-2) \cdots (n-r+1)$$
$$= \frac{n!}{(n-r)!}$$

II. The permutation of n different things taken n at a time is:

$$P(n,n) = n!$$

III. The number of distinct permutations of n things taken all at a time of which p are alike, q others are alike, r others are alike, etc. is:

$$P = \frac{n!}{p!q!r! \cdots}$$

Illustration 1. Find the values of $P(9,2)$, $P(25,3)$, and $P(8,8)$.
Solution.

$$P(9,2) = 9 \cdot 8 = 72.$$
$$P(25,3) = 25 \cdot 24 \cdot 23 = 13800.$$
$$P(8,8) = 8! = 40320.$$

Illustration 2. Nine swimmers compete in a race in which only the first five places receive medals. In how many ways can medals be distributed?
Solution. This is a permutation of 9 things taken 5 at a time.

$$P(9,5) = 9 \cdot 8 \cdot 7 \cdot 6 \cdot 5 = 15120.$$

Illustration 3. How many permutations can be made of the letters in the word *Mississippi* when taken all at a time?

Solution. This is a permutation of 11 things taken all of them at a time. However, some of them are alike; namely 4s, 4i, and 2p. Thus we have:

$$\frac{11!}{4!4!2!} = (11)(10)(9)(7)(5) = 34650.$$

95. Combinations. Another method of choosing a set of objects is to make a selection without regard to the order.

Definition. *All the possible* **selections** *consisting of* **r** *different things chosen from* **n** *given things* $(r \leq n)$, *without regard to the order of selection, are called the* **combinations.**

Symbolically, we write $C(n,r)$ which is read "The number of combinations of n things taken r at a time." The formula is:

$$C(n,r) = \frac{P(n,r)}{r!} = \frac{n!}{r!(n-r)!}$$

The total number of combinations of n things taken successively $1, 2, 3, \ldots, n$ at a time is $2^n - 1$.

Illustration 1. Find the value of $C(6,3)$, $C(26,23)$, and $C(12,8)$.
Solution.

$$C(6,3) = \frac{6 \cdot 5 \cdot 4}{3 \cdot 2 \cdot 1} = 20.$$

$$C(26,23) = \frac{26!}{(3!)(23!)} = \frac{26 \cdot 25 \cdot 24}{3 \cdot 2 \cdot 1} = 2600.$$

$$C(12,8) = \frac{12!}{(4!)(8!)} = \frac{12 \cdot 11 \cdot 10 \cdot 9}{4 \cdot 3 \cdot 2 \cdot 1} = 495.$$

Illustration 2. In how many ways can a hand of thirteen cards be dealt from a deck of fifty-two cards so as to contain exactly 10 spades, if the spades are selected first?

Solution. There are 13 spades in a normal deck from which we must select 10. This leaves 39 cards from which we must select 3. Thus:

$$C(13,10) = \frac{13!}{3!10!} = 286.$$

$$C(39,3) = \frac{39 \cdot 38 \cdot 37}{3 \cdot 2 \cdot 1} = 9139.$$

The number of ways this can be done together is the product:

$$(9139)(286) = 2,613,754.$$

Illustration 3. How many different sums of money can be formed from a penny, a nickel, a dime, and a quarter?

Solution. We have 4 things from which to select successively 1, 2, 3, and 4 at a time. Thus $n = 4$ in the formula:

$$2^n - 1 = 2^4 - 1 = 16 - 1 = 15.$$

96. Elementary Probability. The theory of probability is becoming more and more prominent in the affairs of modern society. Originally developed in connection with games of chance, mortality, and insurance, it is now being applied to complex scientific research and engineering. We shall discuss only the elementary concepts.

Suppose an event can result in one of n ways, each of which, in a single trial, is equally likely. Let s of these ways be considered successes and the others, $n - s$, failures. Then the probability, p, that one particular trial will result in a success is given by:

$$p = \frac{s}{n}.$$

The probability, q, that one trial will result in a failure is given by:

$$q = \frac{n - s}{n}.$$

We note that:

$$p + q = \frac{s}{n} + \frac{n - s}{n} = 1,$$

$$q = 1 - p,$$

$$0 \le p \le 1.$$

If p is the probability of success in one trial of an event, then pk is the **probable number** of successes in k trials. If a person is to receive a sum of S dollars in case a certain event results successfully with a probability p, then the value of his **expectation** is pS dollars.

Illustration 1. What is the probability that one card drawn from a deck of 52 cards will be a spade? an ace? the ace of spades?

Solution. There are 52 cards; therefore $n = 52$.

(a) There are 13 spades; thus $s = 13$ and we have $p = \frac{13}{52} = \frac{1}{4}$.

(b) There are 4 aces; thus $s = 4$ and we have:

$$p = \frac{4}{52} = \frac{1}{13}.$$

(c) There is one ace of spades; thus $s = 1$ and we have:

$$p = \frac{1}{52}.$$

Illustration 2. What is the probability of throwing a seven with a pair of dice?

Solution. Each die is numbered from 1 to 6; thus there are $(6)(6) = 36$ combinations with a pair of dice. A seven can be made with any of the combinations 1 and 6, 2 and 5, 3 and 4, each of which can be made in two ways for a total of 6 successful possibilities. Thus $n = 36$, $s = 6$, and $p = \frac{6}{36} = \frac{1}{6}$.

Illustration 3. If the probability of throwing a seven with a pair of dice is $\frac{1}{6}$, what is the probable number in six tries?

Solution. The probable number is $6(\frac{1}{6}) = 1$.

97. Exercise IX.

1. Find l and s in the following A.P.:

   (a) $d = -3$, $a = 57$, and $n = 6$.
   (b) $\frac{1}{3}, \frac{5}{6}, \frac{4}{3}, \ldots$ to 7 terms.

2. Find l and s in the following G.P.:

   (a) $\frac{1}{2}, \frac{1}{4}, \frac{1}{8}, \ldots$ to 8 terms.
   (b) $\sqrt{2}, \sqrt{6}, 3\sqrt{2}, \ldots$ to 7 terms.

3. Insert seven arithmetic means between 2 and 6.

4. Insert five geometric means between $\frac{1}{2}$ and 2048.

5. Change the repeating decimal 0.3454545 . . . into an equivalent common fraction.

6. Find the sixth term in the H.P. $\frac{1}{3}, \frac{2}{7}, \frac{1}{4}, \ldots$.

7. How many integers, each of four different figures, can be formed from the digits 1, 2, 4, 7, 8, and 9?

8. How many permutations can be made of the letters in the word *trigonometry* when taken all at a time?

9. How many committees of 6 people can be selected from a group of 12 people?

10. In how many ways can four bridge hands be dealt from a deck of cards?

11. If a pair of dice is tossed, what is the probability of throwing a 2? a 3? a 4? a 5? a 6? a 7? an 8? a 9? a 10? an 11? a 12?

12. There is one winning ticket in a box containing 50 tickets and it pays $10.00. What is the value of the expectation of a single draw?

10

THEORY OF EQUATIONS

98. The Rational Integral Equation. We have already discussed the solution of equations and in particular the linear and the quadratic equation. In this chapter we shall discuss higher-degree equations. Consider the equation:

$$(1) \qquad a_0x^n + a_1x^{n-1} + a_2x^{n-2} + \cdots + a_{n-1}x + a_n = 0,$$

where n is a positive integer and a_i $(i = 0,1,\cdots,n)$ is a constant with $a_0 \neq 0$. This equation is called a **rational integral equation of the nth degree in x.** The left member is called a **polynomial of the nth degree in x.** It can also be expressed in functional notation:

$$f(x) = a_0x^n + a_1x^{n-1} + a_2x^{n-2} + \cdots + a_n$$

and is referred to as a **rational integral function of the nth degree in x.**

Illustration 1. Arrange the following equation in standard form and give the values of n and a_i:

$$(2x^2 - 3)^2 = kx^2 - 4x.$$

Solution.

$$(2x^2 - 3)^2 = kx^2 - 4x,$$
$$4x^4 - 12x^2 + 9 - kx^2 + 4x = 0,$$
$$4x^4 - (12 + k)x^2 + 4x + 9 = 0.$$

$n = 4$, $a_0 = 4$, $a_1 = 0$, $a_2 = -12 - k$, $a_3 = 4$, $a_4 = 9$.

Illustration 2. Write the product $\overset{4}{\underset{i=1}{\Pi}} (x - i)$ as a polynomial in x.

Solution. The product indicated is:

$$\overset{4}{\underset{i=1}{\Pi}} (x - i) = (x - 1)(x - 2)(x - 3)(x - 4)$$
$$= x^4 - 10x^3 + 35x^2 - 50x + 24.$$

$n = 4$, $a_0 = 1$, $a_1 = -10$, $a_2 = 35$, $a_3 = -50$, $a_4 = 24$.

155

99. Remainder and Factor Theorems. There are two important theorems to consider before discussing the solutions of a rational integral equation.

Remainder Theorem. *If a polynomial $f(x)$ is divided by $(x - r)$ until a remainder independent of x is obtained, this remainder is equal to $f(r)$.*

Proof. By definition (see Section 7):

$$\text{Dividend} = \text{divisor} \times \text{quotient} + \text{remainder}.$$
$$f(x) = (x - r)Q(x) + R.$$

If the polynomial $f(x)$ is of degree n in x then $Q(x)$ is of degree $(n - 1)$ in x and R is a constant. This definition is true for all values of x and, in particular, for $x = r$; thus:

$$f(r) = (r - r)Q(r) + R.$$

Since $Q(x)$ is a polynomial in x, $Q(r)$ is a number and:

$$(r - r)Q(r) = 0 \times Q(r) = 0.$$
$$\therefore \quad R = f(r).$$

Factor Theorem. *If r is a zero of the polynomial $f(x)$, then $(x - r)$ is a factor of $f(x)$.*

Proof. Let us divide $f(x)$ by $(x - r)$ and employ the remainder theorem to write:

$$f(x) = (x - r)Q(x) + f(r).$$

By hypothesis r is a zero of $f(x)$; i.e., $f(r) = 0$.

$$\therefore \quad f(x) = (x - r)Q(x).$$

We thus see that $(x - r)$ is a factor.

The factor theorem also states that $(x - r)$ is a factor of $f(x)$ if r is a root of the equation $f(x) = 0$.

Illustration 1. Find the remainder when $x^4 + 2x^3 - 3x^2 + 4x - 5$ is divided by $x + 2$.

Solution. Since we are dividing by $x + 2$ we have $r = -2$, and by the remainder theorem:

$$\begin{aligned} R = f(r) &= f(-2) \\ &= (-2)^4 + 2(-2)^3 - 3(-2)^2 + 4(-2) - 5 \\ &= -25. \end{aligned}$$

Illustration 2. Determine whether $(x - 3)$ is a factor of the polynomial $x^4 - 10x^3 + 35x^2 - 50x + 24$.

Solution. We have $r = 3$ and:

$$f(r) = f(3) = (3)^4 - 10(3)^3 + 35(3)^2 - 50(3) + 24$$
$$= 420 - 420 = 0.$$

Hence, by the factor theorem, $x - 3$ is a factor since $f(3) = 0$.

Illustration 3. Determine a and b so that $x^3 - ax^2 + 2x - 3b$ is divisible by $(x - 1)(x - 2)$.

Solution. By the factor theorem $f(1)$ and $f(2)$ must be zero.

$$f(1) = 1 - a + 2 - 3b = 3 - a - 3b = 0.$$
$$f(2) = 8 - 4a + 4 - 3b = 12 - 4a - 3b = 0.$$

The solution of these two simultaneous linear equations is $a = 3$ and $b = 0$.

100. Synthetic Division. The process of dividing a polynomial by a binomial may be considerably shortened by a method called **synthetic division.** To perform a synthetic division, proceed as follows.

1. Arrange $f(x)$ in descending powers of x:

$$f(x) = a_0 x^n + a_1 x^{n-1} + \cdots + a_{n-1} x + a_n.$$

2. Write the coefficients a_i in order on a line. Supply a zero for each missing power of x.

3. To divide by $x - r$ write r at the right on the first line.

4. Complete the following array:

a_0	a_1	a_2	a_3	a_4	$\cdots$	a_{n-1}	a_n	$\underline{\lfloor r}$
	$b_0 r$	$b_1 r$	$b_2 r$	$b_3 r$	$\cdots$	$b_{n-2} r$	$b_{n-1} r$	
b_0	b_1	b_2	b_3	b_4	$\cdots$	b_{n-1}	b_n	

where:

$$b_0 = a_0,$$
$$b_1 = a_1 + b_0 r,$$
$$b_2 = a_2 + b_1 r,$$
$$\cdot \quad \cdot \quad \cdot \quad \cdot \quad \cdot \quad \cdot$$
$$b_i = a_i + b_{i-1} r,$$
$$\cdot \quad \cdot \quad \cdot \quad \cdot \quad \cdot \quad \cdot$$
$$b_n = a_n + b_{n-1} r = f(r) = R.$$

Illustration 1. Divide $4x^3 - 2x^2 + 5x - 7$ by $x - 2$.

Solution.

$$
\begin{array}{rrrr|r}
4 & -2 & 5 & -7 & \underline{2} \\
  & 8 & 12 & 34 & \\
\hline
4 & 6 & 17 & 27 &
\end{array}
$$

$$Q(x) = 4x^2 + 6x + 17;$$
$$R = 27.$$

Illustration 2. Divide $3x^4 - 20x^2 - 12$ by $x + 3$.
Solution.

$$\begin{array}{rrrrr|r}
3 & 0 & -20 & 0 & -12 & \underline{\;-3\;} \\
 & -9 & 27 & -21 & 63 & \\
\hline
3 & -9 & 7 & -21 & 51 &
\end{array}$$

$Q(x) = 3x^3 - 9x^2 + 7x - 21;$
$R = 51.$

101. Fundamental Theorem. The fundamental theorem of algebra will be assumed without proof since the proof is beyond the scope of this book.

Theorem I. *Every rational integral equation $f(x) = 0$ has at least one root.*

Theorem II. *Every rational integral equation of the nth degree has n roots and no more.*

We shall give the proof of Theorem II. Let:

$$f(x) = a_0 x^n + a_1 x^{n-1} + \cdots + a_{n-1} x + a_n = 0, \quad a_0 \neq 0.$$

By Theorem I there is one root, say r_1. By the Factor Theorem:

$$f(x) = (x - r_1)Q_1(x),$$

where $Q_1(x)$ is a polynomial of degree $n - 1$;

$$Q_1(x) = a_0 x^{n-1} + b_1 x^{n-2} + \cdots + b_n.$$

By Theorem I, $Q_1(x) = 0$ has one root, say r_2. By the Factor Theorem:

$$Q_1(x) = (x - r_2)Q_2(x),$$

where $Q_2(x)$ is a polynomial of degree $n - 2$. We continue this process n times to obtain:

$$f(x) = (x - r_1)(x - r_2)(x - r_3) \cdots (x - r_n)Q_n(x),$$

where:

$$Q_n(x) = a_0 x^{n-n} = a_0.$$

Therefore:

$$f(x) = a_0(x - r_1)(x - r_2) \cdots (x - r_n),$$

and we have exactly n roots, $r_1, r_2, \ldots, r_n$. There can be no more, because, suppose there is another, say r. Then:

$$f(r) = a_0(r - r_1)(r - r_2) \cdots (r - r_n).$$

Now if r is different from any of the r_i then none of the factors $(r - r_i)$ are zero, and since $a_0 \neq 0$ we have $f(r) \neq 0$ so that r cannot be a root of $f(x) = 0$.

102. Concerning the Roots. To find the roots of an equation we shall employ the following properties.

I. **Nonreal roots.** *Nonreal roots occur in conjugate pairs;* i.e., *if* $a + bi$ *is a root, then* $a - bi$ *is a root.*

II. **Quadratic Surds.** *If* $a + \sqrt{b}$ *is a root, then* $a - \sqrt{b}$ *is a root.*

III. **Descartes' Rule of Signs.** *The number of* positive *roots of an equation* $f(x) = 0$ *with real coefficients cannot exceed the number of variations in sign in the polynomial* $f(x)$. *The number of* negative *roots cannot exceed the number of variations in sign in* $f(-x)$.

A variation in sign occurs whenever two successive terms in $f(x)$ differ in sign. To find $f(-x)$ simply change the sign of the odd-power terms.

IV. **Rational Roots.** *If a rational number* b/c *is in its lowest terms and is a root of the rational integral equation:*

$$a_0 x^n + a_1 x^{n-1} + \cdots + a_{n-1} x + a_n = 0,$$

with integral coefficients, then b *is a factor of* a_n *and* c *is a factor of* a_0.

To solve an equation we first try to find the maximum number of positive roots and negative roots; then we try to find the rational roots and depress the equation by factoring out the rational roots; finally when we arrive at a quadratic equation we solve it by the quadratic formula.

Illustration 1. Solve the equation:

$$8x^4 - 18x^3 + 2x^2 + 7x - 6 = 0.$$

Solution.

$$f(x): \quad + \quad - \quad + \quad + \quad -$$

There are three variations in sign in $f(x)$.

$$f(-x): \quad + \quad + \quad + \quad - \quad -$$

There is one variation in sign in $f(-x)$.
The factors of $a_n = -6$: $\pm 1, \pm 2, \pm 3, \pm 6$.
The factors of $a_0 = 8$: $\pm 1, \pm 2, \pm 4, \pm 8$.

$\therefore \dfrac{b}{c}$: $\pm 1, \pm 2, \pm 3, \pm 6, \pm \frac{1}{2}, \pm \frac{1}{4}, \pm \frac{1}{8}, \pm \frac{3}{2}, \pm \frac{3}{4}, \pm \frac{3}{8}$.

First, try the integers and seek $f(x) = 0$ by synthetic division.
We find:

$$
\begin{array}{rrrrr|l}
8 & -18 & 2 & 7 & -6 & \underline{2} \\
  & 16 & -4 & -4 & 6 & \\
\hline
8 & -2 & -2 & 3 & \boxed{0} &
\end{array}
$$

The equation is now depressed to the cubic:

$$8x^3 - 2x^2 - 2x + 3 = 0.$$

We find:

$$
\begin{array}{rrrr|r}
8 & -2 & -2 & 3 & \underline{-\tfrac{3}{4}} \\
  & -6 & 6 & -3 & \\
\hline
8 & -8 & 4 & \boxed{0} &
\end{array}
$$

The depressed equation is:

$$2x^2 - 2x + 1 = 0.$$

By quadratic formula $x = \tfrac{1}{2} \pm \tfrac{1}{2}i$.

$\therefore$ Roots are $2, -\tfrac{3}{4}, \tfrac{1}{2} \pm \tfrac{1}{2}i$.

Illustration 2. Solve the equation:

$$6x^4 + x^3 - 26x^2 - 4x + 8 = 0.$$

Solution.

No more than 2 positive roots.

No more than 2 negative roots.

Factors of $a_n = 8$: $\pm 1, \pm 2, \pm 4, \pm 8$.

Factors of $a_0 = 6$: $\pm 1, \pm 2, \pm 3, \pm 6$.

$\therefore \dfrac{b}{c}$: $\pm 1, \pm 2, \pm 4, \pm 8, \pm \tfrac{1}{2}, \pm \tfrac{1}{3}, \pm \tfrac{1}{6}, \pm \tfrac{2}{3}, \pm \tfrac{4}{3}, \pm \tfrac{8}{3}$.

By trial:

$$
\begin{array}{rrrrr|r}
6 & 1 & -26 & -4 & 8 & \underline{2} \\
  & 12 & 26 & 0 & -8 & \\
\hline
6 & 13 & 0 & -4 & \boxed{0} & \qquad \therefore \quad x = 2. \\
  & -12 & -2 & 4 & \underline{-2} & \\
\hline
6 & 1 & -2 & \boxed{0} & & \qquad \therefore \quad x = -2. \\
  & 3 & 2 & \underline{\tfrac{1}{2}} & & \\
\hline
6 & 4 & \boxed{0} & & & \qquad \therefore \quad x = \tfrac{1}{2}.
\end{array}
$$

$6x + 4 = 0;\ x = -\tfrac{2}{3}.$

Roots are: $2, -2, \tfrac{1}{2}, -\tfrac{2}{3}$.

103. Graph of a Polynomial. In order to plot a polynomial we first find a table of values by substituting chosen values of x into $f(x)$. The values of $f(x)$ are most easily obtained by synthetic division. Next, we plot these points on coordinate paper and join the points with a smooth curve.

Illustration 1. Find the graph of:

$$f(x) = 6x^4 + x^3 - 26x^2 - 4x + 8.$$

Solution. Calculate the table of values.

x	-3	-2	-1	0	1	2	3	$-\frac{3}{2}$	$\frac{3}{2}$	$-\frac{2}{3}$	$\frac{1}{2}$
$f(x)$	245	0	-9	8	-15	0	275	$-\frac{35}{2}$	$-\frac{91}{4}$	0	0

Plot these points on a coordinate system with the $f(x)$ units different from the x units. Since the end points are large we use them only to indicate whether $f(x)$ is positive or negative. The values at $x = -\frac{3}{2}$, $\frac{3}{2}$, $-\frac{2}{3}$, $\frac{1}{2}$ were computed as an afterthought to give a smoother curve. Fig. 71 shows the plot.

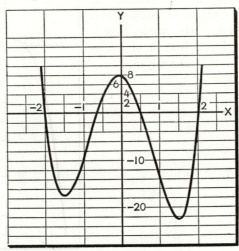

Fɪɢ. 71

Depending upon our interest, this may not be the most rigorous plotting of the graph.* However, at this stage of our study we should be concerned only with the approximate place where the function crosses the X-axis since this gives the zeroes of $f(x)$ or the real roots of $f(x) = 0$.

Illustration 2. Find the graph of $f(x) = x^3 + x^2 + x + 1$.

Solution. Calculate the table of values.

x	-3	-2	-1	0	1	2	3
$f(x)$	-20	-5	0	1	4	15	40

The plot is shown in Fig. 72, p. 162.
We notice that the curve crosses the X-axis in only one point. Since the equation $f(x) = 0$ is a cubic, it has three solutions, but since it crosses

---

* A more rigorous discussion is given in Section 142, D.

the X-axis only once, it has only one real root; the other two are imaginary.

Plotting the function in an equation will aid in finding the roots; if the roots are irrational, it will give a rough first approximation. We shall now turn to a method of finding irrational roots, called Horner's method. Another method will be discussed in Chapter 15, Section 154. More advanced methods may be investigated by studying a branch of mathematics called numerical analysis.

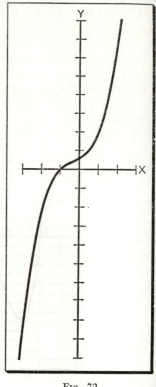

Fig. 72

104. Horner's Method. So far we have considered only the rational roots of an equation. There are many methods for finding the approximate irrational roots. We shall employ a method of successive approximations known as **Horner's method.** Before this method is applied, all rational roots should have been removed, and we need consider only the depressed equation. The method is one of continuous synthetic division which diminishes the roots by a given number.

Illustration 1. Find the real root of:

$$x^3 - x^2 + 4x - 5 = 0.$$

Solution. First plot the function $f(x)$. From the graph it is seen that the function crosses the X-axis between 1 and 2, so we shall first diminish the root by 1. To accomplish this, perform a continuous synthetic division, i.e.:

```
    1  -1    4    -5   | 1
          1    0     4
    ---------------------
    1   0    4 | -1
          1    1 |
    --------------
    1   1 | 5
          1 |
    ------
    1   2
```

The boldfaced numbers are the coefficients of the transformed equation:

$$f_1(x_1) = x_1^3 + 2x_1^2 + 5x_1 - 1 = 0.$$

To obtain the next approximation, ignore the higher-order terms and consider:

$$5x_1 - 1 = 0 \quad \text{or} \quad x_1 = \tfrac{1}{5} = .2.$$

Compute now $f_1(.2)$:

```
    1    2      5       -1       | .2
         .2    .44     1.088
    ------------------------------
    1   2.2   5.44    | +.088 |
```

Since the function was negative (-1) at $x = 1$ and is now positive $(+.088)$ at $x = 1.2$ the trial divisor of .2 is too large; thus we diminish the roots of the equation by .1.

```
    1    2      5       -1       | .1
         .1    .21      .521
    ------------------------------
    1   2.1   5.21 |  -.479
         .1    .22 |
    ----------------
    1   2.2 | 5.43
         .1 |
    --------
    1   2.3
```

The second transformed equation is:

$$f_2(x_2) = x_2^3 + 2.3x_2^2 + 5.43x_2 - .479 = 0.$$

Now solve $5.43x_2 - .479 = 0$ to give $x_2 = .08+$ as a second trial divisor. By synthetic division we find $f_2(.08) = -.029368$ and $f_2(.09) = .029059$. Therefore we diminish the roots by .08.

```
    1   2.3     5.43      -.479      | .08
        .08    .1904     .449632
    ----------------------------------
    1   2.38   5.6204 |  -.029368
        .08    .1968  |
    -------------------
    1   2.46 | 5.8172
        .08  |
    ----------
    1   2.54
```

We now have:

$$f_3(x_3) = x_3{}^3 + 2.54x_3{}^2 + 5.8172x_3 - .029368.$$

The next trial divisor is $.029368 \div 5.8172 = .005+$. By synthetic division:

$$f_3(.005) = -.000218+ \quad \text{and} \quad f_3(.006) = +.0056+.$$

Consequently, the next value of x is .005. We can continue this process as far as desired. The root is $x = \mathbf{1.185+}$.

105. The Cubic Equation. Consider the general cubic equation:

$$x^3 + bx^2 + cx + d = 0. \tag{1}$$

Let:

$$x = y - \tfrac{1}{3}b \tag{2}$$

and substitute into equation (1). After some simplification we obtain:

$$y^3 + py + q = 0 \tag{3}$$

where:

$$p = c - \frac{b^2}{3} \quad \text{and} \quad q = d - \frac{bc}{3} + \frac{2b^3}{27}. \tag{4}$$

Equation (3) is called the **reduced cubic,** and we can see that the transformation removed the x^2 term. We shall now solve equation (3) by letting:

$$y = u + v \quad \text{and} \quad 3uv + p = 0, \tag{5}$$

which imposes two conditions on u and v. A substitution into equation (3) yields:

$$u^3 + v^3 + q = 0,$$

and since:

$$v = -\frac{p}{3u}$$

we finally obtain:

$$u^6 + qu^3 - \frac{p^3}{27} = 0. \tag{6}$$

Equation (6) can be solved for u^3 by the quadratic formula:

$$u^3 = -\frac{q}{2} \pm \sqrt{R} = N \tag{7}$$

where:

$$R = \frac{p^3}{27} + \frac{q^2}{4}. \tag{8}$$

If we choose the "$+$" sign for the radical, we get:

$$v^3 = -\frac{q}{2} - \sqrt{R} = M. \tag{9}$$

The solutions are:

$$u = \sqrt[3]{N}, \quad \omega\sqrt[3]{N}, \quad \omega^2\sqrt[3]{N},$$
$$v = \sqrt[3]{M}, \quad \omega\sqrt[3]{M}, \quad \omega^2\sqrt[3]{M},$$

where:

$$\omega = -\tfrac{1}{2} + \tfrac{1}{2}i\sqrt{3} \quad \text{and} \quad \omega^2 = -\tfrac{1}{2} - \tfrac{1}{2}i\sqrt{3}.$$

The conditions (5) now yield:

$$y_1 = \sqrt[3]{N} + \sqrt[3]{M},$$
$$y_2 = \omega\sqrt[3]{N} + \omega^2\sqrt[3]{M},$$
$$y_3 = \omega^2\sqrt[3]{N} + \omega\sqrt[3]{M},$$

and from equation (2) we finally get:

$$\left.\begin{aligned} x_1 &= \sqrt[3]{N} + \sqrt[3]{M} - \tfrac{1}{3}b, \\ x_2 &= \omega\sqrt[3]{N} + \omega^2\sqrt[3]{M} - \tfrac{1}{3}b, \\ x_3 &= \omega^2\sqrt[3]{N} + \omega\sqrt[3]{M} - \tfrac{1}{3}b. \end{aligned}\right\} \qquad (10)$$

Although the formulas may look somewhat complicated, they are not too difficult to apply. The steps are:

(a) Obtain p and q by formulas (4).
(b) Obtain R by formula (8).
(c) Obtain N and M by formulas (7) and (9).
(d) Obtain x_i ($i = 1,2,3$) by formulas (10).

Illustration. Find the roots of:

$$x^3 + 3x^2 + 6x + 4 = 0.$$

Solution. Calculate the entries in the following table.

b	c	d	p	q	R	N	M
3	6	4	3	0	1	1	-1

Then:
$$x_1 = 1 - 1 - 1 = -1,$$
$$x_2 = \omega(1) + \omega^2(-1) - 1 = -1 + i\sqrt{3},$$
$$x_3 = \omega^2(1) + \omega(-1) - 1 = -1 - i\sqrt{3}.$$

If we can express the roots of an equation by an algebraic formula involving the coefficients of the equation, we say that it can be solved *algebraically*. We have already shown that to be the case for the linear, quadratic, and cubic equations. It is also the case for the fourth-degree equation. However, *rational integral equations of degree higher than the fourth cannot, in general, be solved algebraically.*

106. Exercise X.

1. Prove the Remainder and Factor Theorems.

2. Show that the continuous synthetic division employed in Horner's method yields the coefficients of the equation obtained by replacing x by $x - r$. (Hint. Use the general cubic as the equation.)

3. Find all the roots of:

   (a) $x^4 + 3x^3 - 9x^2 + 3x - 10 = 0.$
   (b) $24x^4 + 2x^3 - 27x^2 + x + 6 = 0.$
   (c) $2x^5 - 9x^4 - 26x^3 + 3x^2 + 6x = 0.$

4. Obtain the graph of 3(b) in the range $-\frac{3}{2} \leq x \leq 1.$

5. Using Horner's method find the indicated root to two decimal places:

   (a) $x^4 - x^3 - 1 = 0$, smallest positive root.
   (b) $2x^4 - 2x - 3 = 0$, smallest negative root.

6. Solve algebraically the cubic equations:

   (a) $x^3 + 6x^2 + 12x + 10 = 0.$
   (b) $x^3 + 6x^2 + 12x + 14 = 0.$

11

LOGARITHMS

107. Definitions. The technological advances of modern times and the complexity of modern business have placed a heavy demand upon the computation branch of mathematics. This demand has resulted in the creation of numerous automatic calculating machines, ranging from the slide rule to the multimillion-dollar "giant brains." Consequently, numerical methods have become more important in recent years. One of the biggest aids to computation is based on the theory of logarithms. To perform calculations by means of logarithms only a table of logarithms and a knowledge of addition is needed. It is still the most practical way for an individual to calculate numerical answers.

Definition. *The* **logarithm** *of a number N to the base* **a** *is the exponent x of the power to which the base must be raised to equal the number N.*

The base *a* must be positive and not equal to one; furthermore, we shall limit ourselves to the case where *N* is positive. Let us employ the notation:

$$\log_a N \equiv \text{the logarithm of } N \text{ to the base } \mathbf{a}.$$

Then the definition states:

If $N = a^x$, *then* $x = \log_a N$.

Under these conditions the logarithm is unique; i.e., every positive number has one and only one logarithm, and every logarithm represents one and only one number.

Illustrations.

Exponential Form	Logarithmic Form
$3^2 = 9$	$\log_3 9 = 2$
$25^{\frac{1}{2}} = 5$	$\log_{25} 5 = \frac{1}{2}$
$2^{-3} = \frac{1}{8}$	$\log_2 \frac{1}{8} = -3$

108. Properties of Logarithms. We recall the following laws of exponents (see Section 6):

I. $a^m a^n = a^{m+n}$.   III. $(a^m)^k = a^{km}$.

II. $\dfrac{a^m}{a^n} = a^{m-n}$.   IV. $\sqrt[q]{a^m} = a^{\frac{m}{q}}$.

Corresponding to these laws of exponents we have four properties of logarithms.

Property I.   *The logarithm of a product is equal to the sum of the logarithms of its factors.*

$$\log_a MN = \log_a M + \log_a N.$$

Proof.

Let: $m = \log_a M$ and $n = \log_a N$.
By definition: $M = a^m$ and $N = a^n$.
Then: $MN = a^m a^n = a^{m+n}$.
By definition: $\log_a MN = m + n = \log_a M + \log_a N$.

This property is true for any number of factors; i.e.:

$$\log_a MNR = \log_a M + \log_a N + \log_a R,$$

etc. for more factors.

Property II.   *The logarithm of a quotient is equal to the logarithm of the numerator minus the logarithm of the denominator.*

$$\log_a \frac{M}{N} = \log_a M - \log_a N.$$

Prove this by using formula (II) of the laws of exponents given above.

Property III.   *The logarithm of the kth power of a number equals k times the logarithm of the number.*

$$\log_a N^k = k \log_a N.$$

Proof.

Let $n = \log_a N$;   then $N = a^n$.
By the laws of exponents: $N^k = (a^n)^k = a^{kn}$.
By definition: $\log_a N^k = kn = k \log_a N$.

Property IV.   *The logarithm of the qth root of a number is equal to the logarithm of the number divided by q.*

$$\log_a \sqrt[q]{N} = \frac{1}{q} \log_a N.$$

Prove this property by the laws of exponents.

Illustration. Express:

$$\log_a \frac{y^2 \sqrt{x}}{zw^3}$$

as the algebraic sum of logarithms.

Solution. By Property II:

$$\log_a \frac{y^2 \sqrt{x}}{zw^3} = \log_a (y^2 \sqrt{x}) - \log_a(zw^3).$$

By Property I: $\qquad = \log_a y^2 + \log_a \sqrt{x} - [\log_a z + \log_a w^3]$

By Props. III, IV: $\qquad = 2 \log_a y + \frac{1}{2} \log_a x - \log_a z - 3 \log_a w.$

Although the base can be any positive number except unity, there are only two values of a that are widely employed. For computational purposes $a = 10$ and the logarithms are then called **common logarithms.** When common logarithms are used, the base is usually omitted in writing and the symbol log N is understood to mean the logarithm to the base 10. The second value of the base is:

$$a = e = 2.71828 \cdots,$$

an irrational number, and the logarithms are called the **natural logarithms.** A special symbol has been adopted for the natural logarithms; thus:

$$\log_e N = \ln N.$$

109. Common Logarithms. It is easy to write down a table of numbers which are integral powers of 10 and the corresponding common logarithms.

Exponential Form	Logarithmic Form
. . . .	. . . .
$10^3 = 1000$	$\log 1000 = 3$
$10^2 = 100$	$\log 100 = 2$
$10^1 = 10$	$\log 10 = 1$
$10^0 = 1$	$\log 1 = 0$
$10^{-1} = 0.1$	$\log 0.1 = -1$
$10^{-2} = 0.01$	$\log 0.01 = -2$
$10^{-3} = 0.001$	$\log 0.001 = -3$
. . . .	. . . .

To find the logarithm of a number which is not an integral power of 10 can best be explained by considering an example such as $N = 110$. Since 110 is between 100 and 1000, it is natural to suppose, from the above table, that log 110 is between 2 and 3 or, in other words, the logarithm is $2 +$ (a proper fraction). Since we can express the proper fraction in decimal form, we have, in general:

$$\log N = (\text{an integer}) + (0 \leq \text{decimal fraction} < 1).$$

The integral part is called the **characteristic**. The decimal fraction is called the **mantissa**. Thus:

$$\log N = \text{characteristic} + \text{mantissa}.$$

Since the mantissas may be nonrepeating infinite decimals, they are approximated to as many places as desired. The approximations have been tabulated in four-place, five-place, or higher-place tables which are called logarithmic tables. Thus the mantissas, or decimal parts, are found from tables and the values in the tables are **always positive**.

The characteristic is determined according to the following two rules:

Rule 1. *If the number N is greater than* 1, *the characteristic of its logarithm is* **one** *less than the number of digits to the left of the decimal point.*

Rule 2. *If the number N is less than* 1, *the characteristic of its logarithm is* **negative**; *if the first digit which is not zero occurs in the kth decimal place, the characteristic is* $-k$.

Since the characteristic and mantissa are combined to give the complete logarithm:

$$\log N = \text{characteristic} + \text{mantissa},$$

and, further, since the mantissa is always positive, it is best to write a negative characteristic, $-k$, as $(10 - k) - 10$.

Illustration 1. Write the logarithms given a mantissa = .3942 and characteristics 1, 0, -1, and -2.

Solution.

Characteristic	Logarithm
1	1.3942
0	0.3942
-1	$9.3942 - 10$
-2	$8.3942 - 10$

Illustration 2. Find the characteristics of the logarithms of the numbers given in the left column.

Solution.

Number	Characteristic
197.3	2
81.72	1
6.291	0
0.3962	9 − 10
0.0815	8 − 10
0.000073	5 − 10

If the logarithm is given, the problem becomes one of finding the number corresponding to this logarithm. The number is called the **antilogarithm.** The characteristic of the given logarithm determines the position of the decimal point in the antilogarithm, and the mantissa determines the digits in the antilogarithm. In placing the decimal point we use the same two rules given for determining the characteristic of a number. However, we must remember it is a reverse problem.

Illustration. The digits of an antilogarithm are 7329. Place the decimal point if the characteristic is 1, 2, 8 − 10.
Solution.

Characteristic	Number
1	73.29
2	732.9
8 − 10 = −2	0.07329

110. Tables and Their Use. To find the mantissa of a logarithm we shall use the Table of the Common Logarithms in the back of this book. This table is a four-place table, and, although it will not yield as accurate results as a table of more places, it will serve to illustrate the methods.* The numbers .04021, .4021, 402100.0 are said to have the *same sequence of digits.* The *mantissa* of the logarithm for each of these numbers is the *same;* the characteristics are, of course, different.

I. To find the logarithm of a given number.

Illustration 1. Find log 32.4.
Solution. By Rule 1 the characteristic is 1. To find the mantissa turn

For a five-place table see Kaj L. Nielsen, *Logarithmic and Trigonometric Tables* (New York: Barnes & Noble, 1943), pp. 2–21.

to the table and locate the first two digits (32) in the left column headed
by "N." In the "32" row and in the column headed by "4" (the third
digit of the number) find 5105. This number is the mantissa.

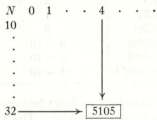

Thus log 32.4 = 1.5105.

Illustration 2. Find log .06732.

Solution. By Rule 2 the characteristic is $8 - 10$. To find the mantissa
it is necessary to interpolate. We shall use simple linear interpolation
as we did in Section 23. Locate "67" in left column. In the "67" row
and column headed by "3" find 8280, and in the column headed by
"4" find 8287. The difference $8287 - 8280 = 7$ is now multiplied by .2
(2 being the fourth significant digit of given number): $(7)(.2) = 1.4$
which is rounded off to the nearest whole number, 1. This number is
added to the smaller mantissa: $8280 + 1 = 8281$. The answer is given by

$$\log .06732 = 8.8281 - 10.$$

II. To find the antilogarithm of a given logarithm.

Illustration 1. Find N if log $N = 7.6503 - 10$.

Solution. First find the mantissa 6503 in the table; it appears in the
column headed by "7" in the row which has "44" at the left under "N."
Thus the sequence of digits is 447.

Now the characteristic is $7 - 10 = -3$. Using Rule 2 for the characteristic, the first significant digit after the decimal point should occur
in the *third* place. The answer is $N = .00447$.

Illustration 2. Find N if log $N = 1.5952$.

Solution. First seek the mantissa in the table. We find 5944 and 5955
corresponding to the numbers 393 and 394. Consequently, our number
is between these two numbers. Form the quotient:

$$\left(\frac{\text{Given mantissa} - \text{smaller table mantissa}}{\text{Larger table mantissa} - \text{smaller table mantissa}} \right) 10.$$

$$10 \left(\frac{5952 - 5944}{5955 - 5944} \right) = \frac{8}{11} (10) = \frac{80}{11} = 7 +.$$

This number is rounded off to an integer which is "attached to" the
smaller number N resulting from the table. Thus 7 is attached to 393 to

give us the sequence of numbers 3937. The given characteristic is 1; so by Rule 1 we have:

$$N = 39.37.$$

111. Logarithms of Trigonometric Functions. It is possible to find the values of the trigonometric functions to a specified accuracy by the use of the trigonometric tables. Thus sin 12° 20′ = .2136. It is therefore possible to find the logarithm of the trigonometric function:

$$\log \sin 12° 20′ = \log 0.2136 = 9.3296 - 10.$$

This procedure, however, would necessitate first using the Table of Natural Trigonometric Functions to find sin 12° 20′ and then using the logarithm table to find log 0.2136. To eliminate such a two-step procedure, tables which give the logarithms of trigonometric functions directly have been computed. An abridged four-place table is given at the back of this book.

The table lists the angles from 0° to 45° in the left column at every 10 minutes. The corresponding logarithms are labeled at the *top* of the pages. The angles from 45° to 90° are given in reverse order in the right-hand column, and the heading of the columns is to be taken from the *bottom* of the pages. In order to aid the interpolation, columns of differences have been calculated. These tabular differences are given in the table in the columns headed by "d" to the right of the columns of "L sin" and "L cos"; "L tan" and "L cot" have a common difference, and the "cd" column between them refers to both.

The sines and cosines are less than one (except for 90° and 0°), and therefore the logarithms of these trigonometric functions have a negative characteristic. Similarly, the logarithms of tan θ for θ < 45° and cot θ for θ between 45° and 90° have a negative characteristic. The first part of the characteristic for the logarithm of these functions is explicitly given in the table; it is, therefore, necessary to *subtract* 10 from the logarithm of the above-mentioned functions as given in the table. The characteristic for the logarithm of tan θ for θ between 45° and 90° and cot θ for θ < 45° as given in the table is the exact characteristic, and no subtraction is to be made in this case. No logarithms are given for the secant and cosecant; if these functions occur in any computation, express them by means of the reciprocal identities in terms of sine and cosine.

Illustrations.

1. Find log sin 23° 43′.

Solution. In the table in the back of the book find 23° 40′ in the *left* column; opposite this number look in the "L sin" column and read 9.6036. In the "d" column between 40′ and 50′ read 29; this is the tabular difference. Take the difference between the given minutes 43 and the tabulated minutes 40 and calculate:

$$\frac{43 - 40}{10}(29) = \frac{3}{10}(29) = 8.7$$

or 9 to the nearest integer. Add this to 9.6036 to get:

$$\log \sin 23° 43′ = \mathbf{9.6045 - 10.}$$

2. Find log cos 61° 36′.

Solution. Find 61° 30′ in the right-hand column of the table, and opposite this number in the column headed by "L cos" at the *bottom* of the page read 9.6787. The tabular difference between 30′ and 40′ is given in the "d" column and reads 24. Calculate:

$$\tfrac{6}{10}(24) = 14.4 \quad \text{or} \quad 14.$$

This amount is to be subtracted from 9.6787 because the cosine is a decreasing function. Thus:

$$\log \cos 61° 36′ = \mathbf{9.6773 - 10.}$$

The student should recall that the sine and tangent functions are increasing functions and that the cosine and cotangent functions are decreasing functions.

Illustrations.

1. Given log tan θ = 9.6932 − 10. Find θ.

Solution. In the "L tan" columns, the number just smaller than 9.6932 is 9.6914 and is found opposite 26° 10′ in the left-hand column. The tabular difference is found from the "cd" column to be 32. Calculate:

$$\frac{9.6932 - 9.6914}{32} = \frac{18}{32} = \frac{9}{16} = .56.$$

This number is now multiplied by 10, which is the tabular difference of the angles, and rounded off to an integer; i.e.:

$$10(.56) = 5.6 = 6 \text{ to the nearest integer.}$$

Since the tangent is an increasing function:

$$\theta = 26° 10′ + 6′ = \mathbf{26° 16′.}$$

2. Given log cot θ = 9.8780 − 10. Find θ.

Solution. In the "L cot" column search for the number closest to 9.8780, and find 9.8797 and 9.8771 in the column headed "L cot" at the bottom of the page and corresponding to the angles 52° 50′ and 53° 0′. Calculate:

$$9.8771 \\ 9.8780 \\ 9.8797 \Big]\Big]\tfrac{9}{26}(10) = [3]$$

to the nearest integer. Subtract 3′ from 53° 0′ to give:

$$\theta = 53° \, 0' - 3' = \mathbf{52° \, 57'}.$$

112. Cologarithms. The cologarithm of a number is defined as the logarithm of the reciprocal of the number:

$$\text{colog } N = \log \frac{1}{N} = \log 1 - \log N = 0 - \log N$$
$$= [10.0000 - 10] - \log N.$$

Illustration. Find colog 35.7.
Solution.

$$\text{colog } 35.7 = \log 1 - \log 35.7$$
$$= 10.0000 - 10$$
$$\text{minus} \qquad \underline{\quad 1.5527 \quad}$$
$$= \;\; 8.4473 - 10.$$

To find a cologarithm mentally:

Subtract each digit of the logarithm of the number, except the last one, from 9 and the last one from 10, and subtract 10 from the result. Thus in the above illustration: $9 - 1 = 8$; $9 - 5 = 4$; $9 - 5 = 4$; $9 - 2 = 7$; $10 - 7 = 3$ giving $8.4473 - 10$.

Cologarithms are used in computation when it appears to be desirable to add all the logarithms instead of subtracting some of them.

Illustration.

$$\log \frac{306}{(98.1)(1.52)} = \log 306 - \log 98.1 - \log 1.52$$
$$= \log 306 + \text{colog } 98.1 + \text{colog } 1.52.$$

(1)	log 306 =	2.4857
(2)	colog 98.1 =	8.0083 − 10
(3)	colog 1.52 =	9.8182 − 10
(4)	sum =	20.3122 − 20

113. Computations Using Logarithms. In carrying out computations using logarithms it is desirable to have a systematic form in which to display the work. We recommend the form given in the illustrations. The student should check each step carefully.

Illustration 1. Find N if $N = \dfrac{5.367}{(12.93)(0.06321)}.$

Solution. We employ logarithms and, using the four properties of logarithms, we get:

$$\log N = \log 5.367 - [\log 12.93 + \log 0.06321].$$

(1)	log 5.367 = 0.7298	
(2)	log 12.93 = 1.1116	
(3)	log 0.06321 = 8.8008 − 10	
(4)	(2) + (3) = 9.9124 − 10	
(5)	(1) − (4) = 0.8174	$N = 6.567$

Note: To subtract (4) from (1) we first change 0.7298 to 10.7298 − 10.

Illustration 2. Find N if $N = \dfrac{(\sqrt[3]{0.9573})(3.21)^2}{98.32}$.

Solution.

$$\log N = \tfrac{1}{3} \log 0.9573 + 2 \log 3.21 - \log 98.32.$$

(1)	$\tfrac{1}{3}$ log 0.9573 = $\tfrac{1}{3}$[9.9810 − 10] = 9.9937 − 10	
(2)	2 log 3.21 = 2[0.5065] = 1.0130	
(3)	(1) + (2) = 11.0067 − 10	
(4)	log 98.32 = 1.9927	
(5)	(3) − (4) = 9.0140 − 10	$N = 0.1033$

Note: To find:

$$\tfrac{1}{3}[9.9810 - 10] = \tfrac{1}{3}[29.9810 - 30] = 9.9937 - 10$$

always rearrange a negative characteristic so that *after* dividing the result will be x.xxxx minus an integer.

Illustration 3. Find N if $N = \sqrt{\dfrac{(0.3592)^3}{673.5}}$.

Solution.

$$\log N = \tfrac{1}{2}[3 \log 0.3592 - \log 673.5].$$

(1)	3 log 0.3592 = 3[9.5553 − 10] = 28.6659 − 30	
(2)	log 673.5 = 2.8284	
(3)	(1) − (2) = 25.8375 − 30	
(4)	$\tfrac{1}{2}$(3) = $\tfrac{1}{2}$[15.8375 − 20] = 7.9188 − 10	
	$N = 0.008294$	

Note: In going from (3) to (4) we made the change

$$25.8375 - 30 = 15.8375 - 20.$$

This was done so that after dividing by 2 we would have x.xxxx − 10.

114. Change of Base. The definition of a logarithm clearly indicates its dependence upon a base. Since different bases may be used, we need a relationship between logarithms to different bases.

Property. *The logarithm of a number* N *to the base* b *is equal to the quotient of its logarithm to the base* a *divided by the logarithm of* b *to the base* a; i.e.:

$$\log_b N = \frac{\log_a N}{\log_a b}.$$

Proof. Let $x = \log_b N$; then by definition:

$$N = b^x.$$

Take the logarithm of both members to the base a.

$$\log_a N = \log_a b^x = x \log_a b.$$

Solve for x:

$$x = \frac{\log_a N}{\log_a b}.$$

Replace x by $\log_b N$:

$$\log_b N = \frac{\log_a N}{\log_a b}.$$

This property is most frequently used to change from the common logarithms to the natural logarithms and vice versa. Thus:

$$\log_{10} N = \frac{\log_e N}{\log_e 10}$$

and

$$\log_e N = \frac{\log_{10} N}{\log_{10} e}.$$

The values of the constants to four decimals are:

$$\log_{10} e = 0.4343 \quad \text{and} \quad \log_e 10 = 2.3026,$$

which are *reciprocals* of each other. Thus we can write:

$$\log_{10} N = 0.4343 \log_e N$$

and

$$\log_e N = 2.3026 \log_{10} N.$$

Illustration 1. Find $\log_9 6$.

Solution.

$$\log_9 6 = \frac{\log_{10} 6}{\log_{10} 9} = \frac{0.7782}{0.9542}$$
$$= 0.8156.$$

Illustration 2. Find $\log_e 14$.
Solution.

$$\log_e 14 = 2.3026 \log_{10} 14 = 2.3026(1.1461)$$
$$= 2.6390.$$

115. Exponential and Logarithmic Equations. The equations which we are about to discuss can be very difficult to solve, and we shall consider only the simple ones.

Definitions. *If the unknown is involved in an equation as an exponent, the equation is called an* **exponential equation.**

If the unknown is involved in an expression whose logarithm is indicated in an equation, the equation is called a **logarithmic equation.**

Examples.

$2^x = 28$ and $3^{x^2} = 5$ are exponential equations.

$\log (x^3 + 2x) = 18$ is a logarithmic equation.

The solution of an exponential equation is usually accomplished by taking the logarithms of both sides and applying the properties of logarithms.

Illustration 1. Solve $5^{x+3} = 625$.
Solution.

$$5^{x+3} = 625 = 5^4.$$

Taking the logarithms of both sides:

$$\log (5^{x+3}) = \log 5^4$$
or
$$(x + 3) \log 5 = 4 \log 5$$
and
$$x + 3 = 4 \quad \text{or} \quad x = 1.$$

Illustration 2. Solve $(2^{x+2})(6^x) = 3^{x+1}$.
Solution.

Take logarithm of each member: $\log[(2^{x+2})(6^x)] = \log (3^{x+1})$.
By properties: $(x + 2) \log 2 + x \log 6 = (x + 1) \log 3$.
Solve for x: $x = \dfrac{\log 3 - 2 \log 2}{\log 2 + \log 6 - \log 3} = \dfrac{\log \frac{3}{4}}{\log 4}$

$$= \frac{9.8751 - 10}{0.6021} = -.2074.$$

Some logarithmic equations can be solved by applying the properties of logarithms and the definition.

Illustration. Solve $\log x + \log (x + 21) = 2$.
Solution.

By Property I: $\qquad\quad \log [x(x + 21)] = 2.$
By definition: $\qquad\quad x^2 + 21x = 10^2 = 100.$
Solving for x: $\qquad\quad x^2 + 21x - 100 = 0.$
$$\qquad\qquad\quad (x + 25)(x - 4) = 0.$$
$$\qquad\qquad\quad x = -25 \quad \text{and} \quad x = 4.$$

Since we have not defined logarithms for negative numbers $x = -25$ is not a solution.

Check: $\qquad\qquad\quad \log 4 + \log 25 = 2.$
$$\qquad\qquad\qquad 0.6021 + 1.3979 = 2.$$

116. Exercise XI.

1. Find the common logarithms of the following numbers: 3.941, 0.00047, 163000.
2. Find the antilogs of $9.8342 - 10$; 2.3469.
3. Find $\log \tan 18° 16'$, $\log \cos 69° 13'$.
4. Find θ if $\log \sin \theta = 9.9341 - 10$.
5. Evaluate to four significant figures by means of logarithms:

(a) $\sqrt{\dfrac{(643.3)(\sqrt{3.288})}{(0.00325)(4325)}}.$

(b) $(73.29)^2(0.009736)(\sqrt{0.3527}).$

6. Find $\ln 2.31$.
7. Solve for x:

(a) $2^{x^2 - 3x} = \frac{1}{2}.$

(b) $\log (x - 5) + \log (3x) = 1.$

12

SOLUTIONS OF TRIANGLES

117. A Triangle. A triangle is a plane figure having six parts: namely, three angles and three sides. If we have given any three parts of which at least one is a side, we can find the unknown parts. This is called the *solution* of the triangle.

Procedure in solving a triangle:

I. *Draw the figure and label it.*

II. *To find any unknown part, use a formula which involves that part but no other unknown. In order to simplify the arithmetic select, whenever possible, a formula which leads to multiplication instead of division.*

III. *As a check, substitute the results in any one of the formulas which was not used in finding the unknown.*

IV. *Use a systematic and neat form in solving triangles.*

The solutions of a triangle are very important, as they find applications in many common problems.

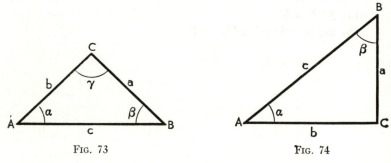

| Fig. 73 | Fig. 74 |

118. Solutions of a Right Triangle. A right triangle is one in which one angle is 90°. Thus one part is already known, and it is only necessary to have given two sides, or an acute angle and a side, in order to solve the triangle.

We recall the formulas (see Chapter 2):

(1) $\sin \alpha = \dfrac{a}{c} = \cos \beta.$ **(2)** $\cos \alpha = \dfrac{b}{c} = \sin \beta.$

(3) $\tan \alpha = \dfrac{a}{b} = \cot \beta.$ **(4)** $\cot \alpha = \dfrac{b}{a} = \tan \beta.$

(5) $a^2 + b^2 = c^2.$ **(6)** $\alpha + \beta = 90°.$

In any of these formulas we can solve for our variable in terms of the others. Thus, for example, from (1) we have:

$$a = c \sin \alpha.$$

Formula (5) can be changed to the forms:

(7) $a = \sqrt{(c - b)(c + b)};$ **(8)** $b = \sqrt{(c - a)(c + a)}.$

These formulas are mainly used in checking the solutions and are readily adapted to logarithmic computation. We cannot stress too strongly the use of a computing form. The student should write down the complete form before looking up any logarithms and doing the arithmetic.

Illustration 1. Solve the right triangle ABC if $\beta = 37° 23'$ and $a = 1.375$.
Solution.

Data: $a = 1.375$; $\beta = 37° 23'.$		
Formulas	Computation	
$\alpha = 90° - \beta$	$\alpha = 90° - 37° 23' =$	$\boxed{52° 37'}$
$\dfrac{b}{a} = \tan \beta,$ or $b = a \tan \beta$	$\log a = \quad 0.1383$ $\log \tan \beta = \quad 9.8832 - 10$	
	$\log b = 10.0215 - 10$	$\boxed{b = 1.051}$
$\dfrac{a}{c} = \cos \beta,$ or $c = \dfrac{a}{\cos \beta}$	$\log a = 10.1383 - 10$ $\log \cos \beta = \quad 9.9001 - 10$	
	$\log c = \quad 0.2382$	$\boxed{c = 1.731}$

Check: $a = \sqrt{(c - b)(c + b)}$ $\log (c - b) = 9.8325 - 10$
$c - b = 0.680$ $\log (c + b) = 0.4443$
$c + b = 2.782$ sum $= 0.2768$
$\log a = 0.1383$ $\longleftarrow$ check $\longrightarrow$ $\frac{1}{2}$ sum $= 0.1384$

Illustration 2. Solve the right triangle ABC if $b = 21.63$ and $c = 47.18$.
Solution.

Data: $b = 21.63$; $c = 47.18$.		
Formulas	Computation	
$\cos \alpha = \dfrac{b}{c}$	$\log b = \quad 1.3351$ $\log c = \quad 1.6737$	
	$\log \cos \alpha = \quad 9.6614 - 10$	$\boxed{\alpha = 62° 42'}$
$\sin \alpha = \dfrac{a}{c}$, or $a = c \sin \alpha$	$\log c = \quad 1.6737$ $\log \sin \alpha = \quad 9.9487 - 10$	
	$\log a = 11.6224 - 10$	$\boxed{a = 41.92}$
$\beta = 90° - \alpha$	$\beta = 90° - 62° 42' = \boxed{27° 18'}$	

Check: $b = \sqrt{(c-a)(c+a)}$ $\log (c-a) = 0.7210$
    $c - a = \ 5.26$ $\log (c+a) = 1.9499$
    $c + a = 89.10$ sum $= 2.6709$
    $\log b = \quad 1.3351$ $\longleftarrow$ ———— ? ———— $\longrightarrow$ $\frac{1}{2}$ sum $= 1.3354$

The check indicates that the agreement is not too good in spite of the fact that there are no computing errors. The disagreement is the result of using four-place tables which yield an answer correct only to ± 1 in the last place, and we must accept this kind of discrepancy. If more extensive tables had been used, we would find that a is closer to 41.93 than to 41.92.

There are many applications of the right triangle and only a few will be discussed here. In solving any problem from descriptive data we suggest the following steps:

I. *Construct a figure.*

II. *Label the figure, introducing single letters to represent unknown angles or lengths.*

III. *Outline the solution and clearly indicate the formulas which are used.*

IV. *Arrange a good computation schematic and perform the arithmetic.*

Let us consider the configuration of Fig. 75, and let O be the point of an observer who is sighting an object C. If the observer is **below** the object, the angle E made by the *line of sight* and the *horizontal* is called the **angle of elevation**. If the observer is **above** the object, the angle D made by the *line of sight* and the *horizontal* is called the **angle of depression**.

Illustration. From the top of a lighthouse 212 feet above a lake, the

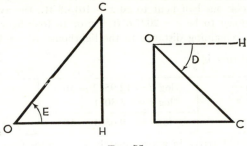

FIG. 75

keeper sights a boat sailing directly towards him. He measures the angle of depression of the boat to be 6° 18′; a half-hour later he measures it to be 13° 10′. Find the distance the boat has sailed between the observations and predict the time when it will arrive at the lighthouse.

Solution. Construct the relative diagram of Fig. 76; i.e., not a scale drawing. B and B' represent the two positions of the boat; LH, the lighthouse; α, first angle of depression; β, second angle of depression. Denote the distances as shown in the figure.

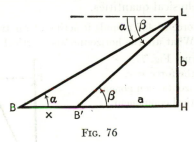

FIG. 76

In $\triangle B'LH$: $b = 212$; $\beta = 13° 10′$.		
$\dfrac{a}{b} = \cot \beta$	$\log b = 2.3263$	
$a = b \cot \beta$	$\log \cot \beta = 0.6309$	
	$\log a = 2.9572$	$a = 906.2$
In $\triangle BLH$: $b = 212$; $\alpha = 6° 18′$.		
$y = x + a$	$\log b = 2.3263$	
$= b \cot \alpha$	$\log \cot \alpha = 0.9570$	
	$\log y = 3.2833$	$y = 1920.0$
$x = y - a = 1920.0 - 906.2 = \boxed{1013.8 \text{ ft.}}$		

Since it took one-half hour to go $x = 1013.8$ ft., the average speed may be assumed to be $s = 2027.6$ ft./hr. or to four figures $s = 2028$ ft./hr. The remaining distance to the lighthouse is $a = 906.2$ ft., so the predicted time is $\dfrac{a}{s}$.

$t = \dfrac{a}{s}$	$\log a = 12.9572 - 10$ $\log s = 3.3071$	
	$\log t = 9.6501 - 10$	$t = .4468$ hr.

The boat should arrive in approximately 27 minutes from the second observation.

A **vector** is a quantity which has magnitude and direction. It is usually represented by an arrow, the length of which represents the magnitude and the head of which indicates its direction. The **sum** of two vectors is called the **resultant** and, in general, is the diagonal of the parallelogram of which the two vectors are the adjacent sides. The two vectors are called the **components** of the resultant along their lines. Vectors are used to represent forces, velocities, accelerations, and other physical quantities.

Illustration. A force of 467 pounds is acting at an angle of 67° 10′ with the horizontal. What are the horizontal and vertical components?

Solution. Construct Fig. 77.
$OR = $ vector of the force $= c$.
$\quad b = OA = $ horizontal component.
$\quad a = OB = $ vertical component.

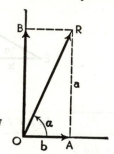

Fig. 77

In $\triangle OAR$: $c = 467$; $\alpha = 67° 10′$.		
$a = c \sin \alpha$	$\log c = 2.6693$ $\log \sin \alpha = 9.9646 - 10$	
	$\log a = 2.6339$	$a = 430.4$ lbs.
$b = c \cos \alpha$	$\log c = 2.6693$ $\log \cos \alpha = 9.5889 - 10$	
	$\log b = 2.2582$	$b = 181.2$ lbs.

119. Oblique Triangles. An **oblique triangle** is a triangle in which there is no right angle. In order to solve the triangle we need to know three parts of which at least one is a side. There are four distinct cases.

Case I.   *Given two angles and a side.*
Case II. *Given two sides and an angle opposite one of them.*
Case III. *Given two sides and the included angle.*
Case IV. *Given three sides.*

To solve the triangle we employ certain laws of the trigonometric functions which we shall discuss in the next sections. We shall also find occasion to use the elementary geometric fact:

$$\textbf{A.} \quad \alpha + \beta + \gamma = 180°.$$

120. Laws of Cosines, Sines, and Tangents. We shall first state the laws and focus the resulting formulas; then a proof will be developed for each of the laws.

The Law of Cosines. *In any triangle, the square of any side is equal to the sum of the squares of the other sides minus twice their product times the cosine of their included angle.*

$$\textbf{B.} \quad \begin{cases} a^2 = b^2 + c^2 - 2bc \cos \alpha; \\ b^2 = a^2 + c^2 - 2ac \cos \beta; \\ c^2 = a^2 + b^2 - 2ab \cos \gamma. \end{cases}$$

We can solve these formulas for the cosines of the angles.

$$\textbf{C.} \quad \begin{cases} \cos \alpha = \dfrac{b^2 + c^2 - a^2}{2bc}; \\[2mm] \cos \beta = \dfrac{a^2 + c^2 - b^2}{2ac}; \\[2mm] \cos \gamma = \dfrac{a^2 + b^2 - c^2}{2ab}. \end{cases}$$

The Law of Sines. *In any triangle, any two sides are proportional to the sines of the opposite angles.*

$$\textbf{D.} \quad \frac{a}{\sin \alpha} = \frac{b}{\sin \beta} = \frac{c}{\sin \gamma}.$$

The Law of Tangents. *In any triangle the difference of any two sides divided by their sum equals the tangent of one-half the difference of the opposite angles divided by the tangent of one-half their sum.*

$$\text{E.} \begin{cases} \dfrac{a-b}{a+b} = \dfrac{\tan\frac{1}{2}(\alpha-\beta)}{\tan\frac{1}{2}(\alpha+\beta)}; \\[2mm] \dfrac{a-c}{a+c} = \dfrac{\tan\frac{1}{2}(\alpha-\gamma)}{\tan\frac{1}{2}(\alpha+\gamma)}; \\[2mm] \dfrac{b-c}{b+c} = \dfrac{\tan\frac{1}{2}(\beta-\gamma)}{\tan\frac{1}{2}(\beta+\gamma)}. \end{cases}$$

The difference of any two sides may be taken in any order, but the difference of the corresponding angles must be in the same order, i.e., if we use $c - b$, then we must use $\gamma - \beta$. Always take the order in which the difference of the sides will be positive.

Proofs. Let a be any side and let β be an acute angle of the triangle. Drop a perpendicular from C to AB, or AB extended. If α is acute we obtain Fig. 78; if α is obtuse we obtain Fig. 79. Let:

$$AD = m, \quad BD = n, \quad \text{and} \quad DC = h.$$

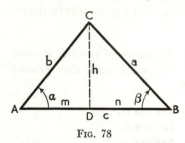

Fig. 78

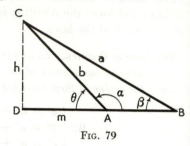

Fig. 79

In either figure we have the following properties by the Pythagorean theorem:

In rt. $\triangle ADC$: $b^2 = h^2 + m^2$; $h^2 = b^2 - m^2$.   (1)

In rt. $\triangle BDC$: $a^2 = h^2 + n^2$.   (2)

Substituting (1) into (2): $a^2 = b^2 - m^2 + n^2$.   (3)

The Law of Cosines.

1. If α is acute, $n = c - m$.

    In $\triangle ADC$: $m = b \cos \alpha$.

    Substituting into (3) we obtain:

$$\begin{aligned} a^2 &= b^2 - m^2 + (c - m)^2 \\ &= b^2 - m^2 + c^2 - 2cm + m^2 \\ &= b^2 + c^2 - 2bc \cos \alpha. \end{aligned}$$

2. If α is obtuse, $n = c + m$.

    In $\triangle ADC$: $m = b \cos \theta$.

    Since $\alpha = 180° - \theta$, $\cos \alpha = -\cos \theta$.

$\therefore$ $m = -b \cos \alpha.$

Substituting in (3):

$$a^2 = b^2 - m^2 + (c + m)^2$$
$$= b^2 - m^2 + c^2 + 2cm + m^2$$
$$= b^2 + c^2 - 2bc \cos \alpha.$$

The Law of Sines.

1. If α is acute:

   In rt. $\triangle ADC$: $h = b \sin \alpha.$

   In rt. $\triangle BDC$: $h = a \sin \beta.$

   By substitution: $a \sin \beta = b \sin \alpha.$

   Dividing both sides by $\sin \alpha \sin \beta$: $\dfrac{a}{\sin \alpha} = \dfrac{b}{\sin \beta}.$

2. If α is obtuse:

   In rt. $\triangle ADC$: $h = b \sin \theta = b \sin \alpha.$

   In rt. $\triangle BDC$: $h = a \sin \beta.$

   Again we have: $a \sin \beta = b \sin \alpha$

   or $\dfrac{a}{\sin \alpha} = \dfrac{b}{\sin \beta}.$

The Law of Tangents.

From the law of sines: $\dfrac{a}{c} = \dfrac{\sin \alpha}{\sin \gamma}.$

Subtracting 1 from both sides: $\dfrac{a}{c} - 1 = \dfrac{\sin \alpha}{\sin \gamma} - 1$

or $\dfrac{a - c}{c} = \dfrac{\sin \alpha - \sin \gamma}{\sin \gamma}.$ (1)

Adding 1 to both sides: $\dfrac{a + c}{c} = \dfrac{\sin \alpha + \sin \gamma}{\sin \gamma}.$ (2)

Dividing each side of (1) by the corresponding side of (2):

$$\frac{a - c}{a + c} = \frac{\sin \alpha - \sin \gamma}{\sin \alpha + \sin \gamma}.$$

Using (22) and (21) of Section 34:

$$\frac{a - c}{a + c} = \frac{2 \cos \frac{1}{2}(\alpha + \gamma) \sin \frac{1}{2}(\alpha - \gamma)}{2 \sin \frac{1}{2}(\alpha + \gamma) \cos \frac{1}{2}(\alpha - \gamma)} = \frac{\tan \frac{1}{2}(\alpha - \gamma)}{\tan \frac{1}{2}(\alpha + \gamma)}.$$

121. The Half-Angle Formulas. A set of formulas known as the half-angle formulas is frequently used in solving a triangle. Let us first define two quantities:

s   denotes one-half the perimeter of the triangle;
r   denotes the radius of the inscribed circle.

Then in any triangle ABC we have:

$$s = \tfrac{1}{2}(a + b + c) \quad \text{and} \quad r = \sqrt{\frac{(s-a)(s-b)(s-c)}{s}}.$$

The first of these two formulas is the direct result of the definition. The second requires a derivation. The easiest proof is the result of the formula for the area of a triangle. Let K denote the area of $\triangle ABC$. See Fig. 80.

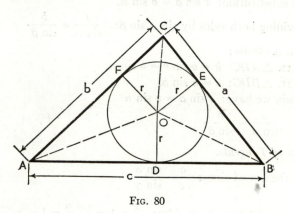

Fɪɢ. 80

Then from elementary geometry we have:

$$K = \text{area of } \triangle AOB + \text{area of } \triangle BOC + \text{area of } \triangle AOC$$
$$= \tfrac{1}{2}rc + \tfrac{1}{2}ra + \tfrac{1}{2}rb = r \cdot \tfrac{1}{2}(a + b + c)$$
$$= rs.$$

It may be recalled from elementary geometry (we shall also prove it using trigonometry; see Section 122) that:

$$K = \sqrt{s(s-a)(s-b)(s-c)}.$$

Thus:

$$rs = \sqrt{s(s-a)(s-b)(s-c)}$$

or dividing by s:

$$r = \sqrt{\frac{s(s-a)(s-b)(s-c)}{s^2}} = \sqrt{\frac{(s-a)(s-b)(s-c)}{s}}.$$

Tangents of the Half-Angles. In any triangle ABC:

F. $\qquad \tan\dfrac{\alpha}{2} = \dfrac{r}{s-a}; \quad \tan\dfrac{\beta}{2} = \dfrac{r}{s-b}; \quad \tan\dfrac{\gamma}{2} = \dfrac{r}{s-c}.$

Proof. From $\triangle ADO$ in Fig. 80:

$$\tan\frac{\alpha}{2} = \frac{\overline{OD}}{\overline{AD}} = \frac{r}{\overline{AD}}.$$

The perimeter of $\triangle ABC$ is:

$$2s = (\overline{AD} + \overline{AF}) + (\overline{BD} + \overline{BE}) + (\overline{CE} + \overline{CF})$$
$$= 2\overline{AD} + 2\overline{BE} + 2\overline{CE}$$

since $\overline{AD} = \overline{AF}$, $\overline{BD} = \overline{BE}$, and $\overline{CE} = \overline{CF}$. Thus:

$$s = \overline{AD} + (\overline{BE} + \overline{CE}) = \overline{AD} + a$$

or $\qquad\qquad \overline{AD} = s - a.$

Therefore:

$$\tan\frac{\alpha}{2} = \frac{r}{s-a}.$$

Similarly for $\tan\dfrac{\beta}{2}$ and $\tan\dfrac{\gamma}{2}$.

122. The Area of the Oblique Triangle. We can solve for the area of the oblique triangle in the following cases:

Case I. *Given two angles and a side.*
Case III. *Given two sides and the included angle.*
Case IV. *Given three sides.*

To find the area for Case II of Section 119 it is necessary to obtain one of the other elements of the triangle first and then use the formulas we shall now derive.

The area in terms of two sides and the included angle is given by:

(1) $K = \frac{1}{2}bc \sin \alpha; \quad K = \frac{1}{2}ac \sin \beta; \quad K = \frac{1}{2}ab \sin \gamma.$

Proof. Construct Fig. 81.

Let b and c be the given sides;
α, the included angle;
h, the altitude.

In $\triangle ACD$: $h = b \sin \alpha$.
Then $K = \frac{1}{2}hc = \frac{1}{2}bc \sin \alpha.$
Similarly for the other formulas.

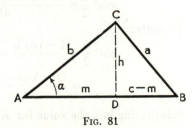

Fig. 81

The area in terms of one side and the angles is given by:

(2) $K = \dfrac{a^2 \sin \beta \sin \gamma}{2 \sin \alpha} = \dfrac{b^2 \sin \alpha \sin \gamma}{2 \sin \beta} = \dfrac{c^2 \sin \alpha \sin \beta}{2 \sin \gamma}.$

Proof. From the law of sines, we have:

$$\frac{c}{\sin \gamma} = \frac{b}{\sin \beta} \quad \text{or} \quad c = \frac{b \sin \gamma}{\sin \beta}.$$

Substituting for c into (1) we get:

$$K = \tfrac{1}{2}bc \sin \alpha = \tfrac{1}{2}b \left(\frac{b \sin \gamma}{\sin \beta} \right) \sin \alpha = \frac{b^2 \sin \alpha \sin \gamma}{2 \sin \beta}.$$

Similarly for the other formulas.

The area in terms of the three sides is given by:

(3) $K = \sqrt{s(s - a)(s - b)(s - c)},$

where:

$$s = \tfrac{1}{2}(a + b + c).$$

Proof. By Section 34:

$$\sin \alpha = 2 \sin \frac{\alpha}{2} \cos \frac{\alpha}{2},$$

so that in (1) we have:

$$K = bc \sin \frac{\alpha}{2} \cos \frac{\alpha}{2}.$$

By Section 34: $2 \cos^2 \tfrac{1}{2}\alpha = 1 + \cos \alpha.$

By Section 120: $\cos \alpha = \dfrac{b^2 + c^2 - a^2}{2bc}.$

Thus:

$$2 \cos^2 \frac{\alpha}{2} = 1 + \frac{b^2 + c^2 - a^2}{2bc} = \frac{(b + c + a)(b + c - a)}{2bc}$$

or

$$\cos^2 \frac{\alpha}{2} = \frac{(b + c + a)(b + c - a)}{4bc} = \frac{s(s - a)}{bc}.$$

Similarly:

$$\sin^2 \frac{\alpha}{2} = \frac{1 - \cos \alpha}{2} = \frac{(a - b + c)(a + b - c)}{4bc}$$

$$= \frac{(s - b)(s - c)}{bc}.$$

Substituting into the value for K, we have:

$$K = bc \sqrt{\frac{s(s-a)}{bc}} \sqrt{\frac{(s-b)(s-c)}{bc}}$$
$$= \sqrt{s(s-a)(s-b)(s-c)}.$$

123. The Ambiguous Case. If two sides and an angle opposite one of them are given (Case II), there may exist two triangles, one triangle, or no triangle. Consequently, Case II is called the ambiguous case. Let the given parts be a, b, and α; then the possibilities are shown in Figs. 82–87.

These conditions which are shown in the figures may be summarized as follows:

If $\alpha < 90°$ $\begin{cases} a < b \sin \alpha; \ \textbf{no } solution. \\ a = b \sin \alpha; \ \textbf{one } solution; \ \beta = 90°. \\ a > b \sin \alpha; \ a < b; \ \textbf{two } solutions. \\ a \geq b; \ \textbf{one } solution. \end{cases}$

If $\alpha \geq 90°$ $\begin{cases} a > b; \ \textbf{one } solution. \\ a \leq b; \ \textbf{no } solution. \end{cases}$

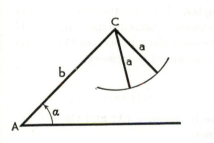

FIG. 82

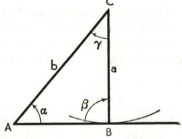

FIG. 83

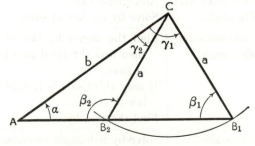

FIG. 84

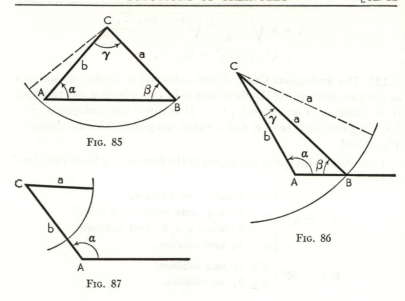

FIG. 85

FIG. 86

FIG. 87

124. Solutions of Oblique Triangles. The laws and formulas which we have been discussing in the previous sections are used to solve the oblique triangles. Referring to the cases given in Section 119 we now give a summary which should be committed to memory.

CASES	SOLUTION
I. Given two angles and a side.	Solve by formula (A) and the law of sines. Find the area by (2).
II. Given two sides and an opposite angle.	Ambiguous case. Solve by the law of sines.
III. Given two sides and the included angle.	Find the angles by law of tangents; then find the third side by the law of sines. If only third side is required, use the law of cosines. Find the area by (1).
IV. Given three sides.	Solve by half-angle formulas. Law of cosines may be used. Find the area by (3).

To illustrate the method of solution we shall work one problem for each case. The student should review the procedure to be used in solving a triangle as stated on page 180. Additional logical steps for the solution of oblique triangles are:

I. Determine the case from the given data.

II. Clearly indicate the unknown parts.

III. Choose the most advantageous formulas as indicated in the summary chart.

IV. Set up the computational procedure with the formulas solved for the unknowns of the given problem.

V. Always include a check.

The use of a neat, systematic computational form cannot be over-emphasized.

Illustration. *Case I.* Solve the triangle ABC if $\alpha = 39° 28'$; $\gamma = 110° 43'$; $a = 36.48$.

Solution.

Data: $\alpha = 39° 28'$; $\gamma = 110° 43'$; $a = 36.48$.		
Formulas	Computation	
$\beta = 180° - (\alpha + \gamma)$	$\beta = 180° - 150° 11' = \boxed{29° 49'}$	
$\dfrac{b}{\sin \beta} = \dfrac{a}{\sin \alpha}$ $b = \dfrac{a \sin \beta}{\sin \alpha}$	log a = 1.5620 log sin β = 9.6966 − 10	
	sum = 11.2586 − 10 log sin α = 9.8032 − 10	
	log b = 1.4554	$\boxed{b = 28.54}$
$\dfrac{c}{\sin \gamma} = \dfrac{a}{\sin \alpha}$ $c = \dfrac{a \sin \gamma}{\sin \alpha}$	log a = 1.5620 log sin γ = 9.9710 − 10	
	sum = 11.5330 − 10 log sin α = 9.8032 − 10	
	log c = 1.7298	$\boxed{c = 53.68}$
$K = \dfrac{a^2 \sin \beta \sin \gamma}{2 \sin \alpha}$	2 log a = 3.1240 log sin β = 9.6966 − 10 log sin γ = 9.9710 − 10	
	sum = 12.7916 − 10	
	log 2 = 0.3010 log sin α = 9.8032 − 10	
	sum = 10.1042 − 10	
	log K = 2.6874	$\boxed{K = 486.9}$

Illustration. *Case II.* Solve the triangle ABC if $c = .513$; $b = .753$; $\gamma = 36° 50'$.

Solution.

Data: $c = .513$, $b = .753$, $\gamma = 36° 50'$.	
Formulas	Computation
$\sin \beta = \dfrac{b \sin \gamma}{c}$ $\beta_2 = 180° - \beta_1$	$\log b = \ \ 9.8768 - 10$ $\log \sin \gamma = \ \ 9.7778 - 10$ $\text{sum} = 19.6546 - 20$ $\quad$ $\boxed{\beta_1 = 61° 39'}$ $\log c = \ \ 9.7101 - 10$ $\log \sin \beta = \ \ 9.9445 - 10$ $\quad$ $\boxed{\beta_2 = 118° 21'}$
$\alpha = 180° - (\gamma + \beta)$	$\alpha_1 = 180° - 98° 29' = \quad \boxed{81° 31'}$ $\alpha_2 = 180° - 155° 11' = \quad \boxed{24° 49'}$
$a_1 = \dfrac{c \sin \alpha_1}{\sin \gamma}$	$\log c = \ \ 9.7101 - 10$ $\log \sin \alpha_1 = \ \ 9.9952 - 10$ $\text{sum} = 19.7053 - 20$ $\log \sin \gamma = \ \ 9.7778 - 10$ $\log a_1 = \ \ 9.9275 - 10$ $\quad$ $\boxed{a_1 = .8462}$
$a_2 = \dfrac{c \sin \alpha_2}{\sin \gamma}$	$\log c = \ \ 9.7101 - 10$ $\log \sin \alpha_2 = \ \ 9.6229 - 10$ $\text{sum} = 19.3330 - 20$ $\log \sin \gamma = \ \ 9.7778 - 10$ $\log a_2 = \ \ 9.5552 - 10$ $\quad$ $\boxed{a_2 = .3591}$
$K_1 = \frac{1}{2}bc \sin \alpha$ $S = \frac{1}{2}bc$	$\log b = \ \ 9.8768 - 10$ $\log c = \ \ 9.7101 - 10$ $\text{colog } 2 = \ \ 9.6990 - 10$ $\log S = \ \ 9.2859 - 10$ $\log \sin \alpha_1 = \ \ 9.9952 - 10$ $\log K_1 = \ \ 9.2811 - 10$ $\quad$ $\boxed{K_1 = .1910}$
$K_2 = \frac{1}{2}bc \sin \alpha_2$	$\log S = \ \ 9.2859 - 10$ $\log \sin \alpha_2 = \ \ 9.6229 - 10$ $\log K_2 = \ \ 8.9088 - 10$ $\quad$ $\boxed{K_2 = .08106}$

Illustration. *Case III.* Solve the triangle ABC if $b = .356$; $c = .938$; $\alpha = 62° 40'$.

Solution. Note first that $\gamma = \frac{1}{2}(\gamma + \beta) + \frac{1}{2}(\gamma - \beta)$; $\beta = \frac{1}{2}(\gamma + \beta) - \frac{1}{2}(\gamma - \beta)$.

Data: $b = .356$; $c = .938$; $\alpha = 62° 40'$.	
Formulas	Computation
$\frac{1}{2}(\gamma + \beta) = \frac{1}{2}(180° - \alpha)$	$\frac{1}{2}(\gamma + \beta) = \frac{1}{2}(180° - 62° 40') = \quad 58° 40'$
$\tan \frac{1}{2}(\gamma - \beta)$ $= \dfrac{(c - b) \tan \frac{1}{2}(\gamma + \beta)}{c + b}$	$\log (c - b) = 9.7649 - 10$ $\log \tan \frac{1}{2}(\gamma + \beta) = 0.2155$ $\text{colog} (c + b) = 9.8881 - 10$
$c = 0.938$ $b = 0.356$ $c - b = 0.582$ $c + b = 1.294$	$\log \tan \frac{1}{2}(\gamma - \beta) = 9.8685 - 10$ $\frac{1}{2}(\gamma - \beta) = 36° 27'$ $\boxed{\gamma = 95° 07'}$ $\boxed{\beta = 22° 13'}$
$a = \dfrac{b \sin \alpha}{\sin \beta}$	$\log b = 9.5514 - 10$ $\log \sin \alpha = 9.9486 - 10$ $\text{colog} \sin \beta = 0.4224$ $\log a = 9.9224 - 10 \qquad \boxed{a = 0.8364}$

Check:

$\log a = 9.9224 - 10$	$\log b = 9.5514 - 10$	$\log c = 9.9722 - 10$
$\log \sin \alpha = 9.9486 - 10$	$\log \sin \beta = 9.5776 - 10$	$\log \sin \gamma = 9.9982 - 10$
Diff. $= 9.9738 - 10$	Diff. $= 9.9738 - 10$	Diff. $= 9.9740 - 10$

The finding of the area is left to the student.

Illustration. *Case IV.* Solve the triangle ABC if $a = 163.6$; $b = 397.5$; $c = 253.7$.
Solution.

Data: $a = 163.6$; $b = 397.5$; $c = 253.7$.	
Formulas	Computation
$s = \frac{1}{2}(a + b + c)$	$a = 163.6$ $b = 397.5$ $c = 253.7$ $2s = 814.8$ $s = 407.4 \longleftarrow$ check $\longrightarrow$ $\quad s - a = 243.8$ $\quad s - b = \quad 9.9$ $\quad s - c = 153.7$ $\quad s = 407.4$
$r^2 = \dfrac{(s - a)(s - b)(s - c)}{s}$	$\log (s - a) = \quad 2.3870$ $\log (s - b) = \quad 0.9956$ $\log (s - c) = \quad 2.1867$ $\log N = \quad 5.5693$ $\log s = \quad 2.6100$ $2 \log r = \quad 2.9593$ $\log r = \quad 1.4796$

$\tan \frac{1}{2}\alpha = \dfrac{r}{s-a}$	$\log r = 11.4796 - 10$ $\log (s-a) = 2.3870$	
	$\log \tan \frac{1}{2}\alpha = 9.0926 - 10$ $\frac{1}{2}\alpha = 7°\ 3'$	$\alpha = 14°\ 6'$
$\tan \frac{1}{2}\beta = \dfrac{r}{s-b}$	$\log r = 1.4796$ $\log (s-b) = 0.9956$	
	$\log \tan \frac{1}{2}\beta = 0.4840$ $\frac{1}{2}\beta = 71°\ 50'$	$\beta = 143°\ 40'$
$\tan \frac{1}{2}\gamma = \dfrac{r}{s-c}$	$\log r = 11.4796 - 10$ $\log (s-c) = 2.1867$	
	$\log \tan \frac{1}{2}\gamma = 9.2929 - 10$ $\frac{1}{2}\gamma = 11°\ 6'$	$\gamma = 22°\ 12'$
Check: $\alpha + \beta + \gamma = 180°$		sum $= 179°\ 58'$
$K = \sqrt{s(s-a)(s-b)(s-c)}$	$\log s = 2.6100$ $\log N = 5.5693$	
$N = (s-a)(s-b)(s-c)$	$2\log K = 8.1793$ $\log K = 4.0896$	$K = 12290$

The use of four-place tables does not permit any greater accuracy than the check above indicates.

125. Some Applications. In this section we consider a few applications.

1. Vectors. The resultant of two vectors is in general the diagonal of a parallelogram of which the two vectors form adjoining sides (see Section 118).

Illustration. Two forces of 50 lbs. and 30 lbs. have an included angle of 60°. Find the magnitude and direction of their resultant.

Solution. Construct the parallelogram and label it as in Fig. 88.

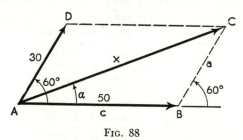

Fig. 88

Since $\overline{AD}$ is parallel to $\overline{BC}$ we have:
$$\angle ABC = \beta = 180° - 60° = 120°.$$

By the law of cosines:
$$x^2 = c^2 + a^2 - 2ac \cos \beta$$
$$= 2500 + 900 - 2(50)(30)(-\tfrac{1}{2})$$
$$= 2500 + 900 + 1500 = 4900.$$

$$\boxed{x = 70 \text{ lbs.}}$$

$$\cos \alpha = \frac{x^2 + c^2 - a^2}{2xc} = \frac{4900 + 2500 - 900}{2(70)(50)} = \frac{13}{14} = .9286.$$

$$\boxed{\alpha = 21° 47'}.$$

2. Navigation. The subject of determining the positions and courses of ships at sea or airplanes in flight is called **navigation**. As a rough check, or if the distances are not too great, we may consider the surface of the earth to be flat and apply the theory of **plane sailing**. In this case the problems reduce to the solution of triangles. As in any subject matter we must first become familiar with the terminology.

The direction from a given point O to a point B in the horizontal plane is described by *the angle through which ON* (the north direction) *must be rotated clockwise in order to coincide with OB*, and this angle is called the **azimuth** of B from O. See Fig. 89.

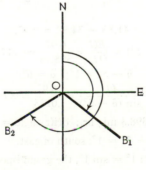

Fig. 89

The **heading** of an airplane is the direction in which the *nose of the airplane is pointing*. The point on the ground directly (vertically) under the airplane describes a path called the **track**, and its *direction* is called the **course** of the airplane. The airplane is in motion with respect to the air, which in turn may be in motion because of the wind.

The speed of the airplane with respect to the air is called the **airspeed** and is in the direction of the heading. Since the wind has a speed in a certain direction there is a resultant of the two speeds with respect to the ground, and this is called the **groundspeed.** If we represent these speeds by vectors and there is an angle between them, this angle is called the **drift angle** and depends upon the *heading* as well as the *airspeed*. This is the familiar concept that the wind blows the airplane off its heading by an amount equal to the drift angle and is true also of sailboats in water.

> **Illustration.** An airplane is heading 5° north of directly east with an airspeed of exactly 400 miles per hour. There is a wind blowing directly from the north with a speed of 41.85 miles per hour. Find the groundspeed, the drift angle, and the course of the airplane.
>
> *Solution.* Construct Fig. 90.

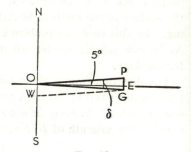

$OP \equiv$ airspeed = 400.
$OW \equiv$ wind velocity = 41.85 = PG.
$OG \equiv$ resultant groundspeed = x.
$\delta \equiv$ drift angle.

In rt. $\triangle OEP$:

$EP = 400 \sin 5°$
    $= 400(.0872) = 34.88.$
$OE = 400 \cos 5°$
    $= 400(.9962) = 398.48.$

In rt. $\triangle OEG$:

$$EG = 41.85 - EP = 41.85 - 34.88 = 6.97.$$

$$\tan (\delta - 5°) = \frac{EG}{OE} = \frac{6.97}{398.48} = .0175.$$

$$\delta - 5° = 1°; \quad \delta = 6°.$$

$$x = \frac{EG}{\sin (\delta - 5°)} = \frac{6.97}{.0175} = 398.3.$$

∴   Groundspeed = 398.3 mi./hr.; drift angle = $\delta = 6°$.

     Course = $\delta - 5° = 1°$, south of east.

We note that since $\tan 1° = \sin 1°$, the groundspeed and the east component are equal.

3. Surveying. The classical application of triangles is in the subject of surveying. For the short distances involved, the earth is assumed to be flat and the problems are reduced to plane figures in two dimensions. We shall illustrate with an example.

Illustration. A surveyor desires to run a straight line due east past an obstruction. See Fig. 91. He measures AB = 780 ft., S 25° 20′ E (25° 30′ east of south) and then runs BC in the direction N 45° 50′ E. Find the length of BC so that C is due east of A, and find the length of AC.

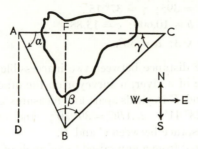

FIG. 91

Solution.

Data: $\angle DAB = 25° 20'$; $AB = 780$ ft.; $\angle FBC = 45° 50'$.		
$\alpha = 90° - \angle DAB = 90° - 25° 20' = 64° 40'$ $\gamma = 90° - \angle FBC = 90° - 45° 50' = 44° 10'$ $\beta = 180° - (\alpha + \gamma) = 180° - 108° 50' = 71° 10'$		
In $\triangle ABC$:	$\log AB = \quad 2.8921$ $\log \sin \alpha = \quad 9.9561 - 10$	
$BC = \dfrac{AB \sin \alpha}{\sin \gamma}$	$\text{sum} = \quad 12.8482 - 10$ $\log \sin \gamma = \quad 9.8431 - 10$	
	$\log BC = \quad 3.0051$	$BC = 1012$ ft.
	$\log AB = \quad 2.8921$ $\log \sin \beta = \quad 9.9761 - 10$	
$AC = \dfrac{AB \sin \beta}{\sin \gamma}$	$\text{sum} = \quad 12.8682 - 10$ $\log \sin \gamma = \quad 9.8431 - 10$	
	$\log AC = 3.0251$	$AC = 1060$ ft.

126. Exercise XII.

Solve the right $\triangle ABC$ and check:

1. $a = 13.74$; $\alpha = 21° 53'$.
2. $b = 6.397$; $\alpha = 57° 17'$.
3. $c = .372$; $\alpha = 40° 10'$.
4. $a = 113.5$; $b = 100.8$.
5. $c = 93.60$; $a = 45.03$.

Solve the following triangles and find the area:

6. $a = 598.4$; $\alpha = 66° 39'$; $\beta = 69° 31'$.

7. $a = .3895$; $c = .5293$; $\beta = 62° 12'$.

8. $c = 270$; $a = 395$; $\gamma = 37° 15'$.

9. $a = 9.782$; $b = 10.600$; $c = 13.88$.

10. $b = 4$; $c = \sqrt{3}$; $\alpha = 30°$.

11. To find the distance between two inaccessible points C and D on one side of a river, a surveyor on the other side picks two points A and B 16 yards apart and measures $\angle CAD = 32° 19'$; $\angle DAB = 28° 41'$; $\angle ABC = 39° 30'$; and $\angle CBD = 36° 20'$. Find the distance between C and D.

12. A shell is fired from a gun raised to an angle of $37° 43'$ with the horizontal. The muzzle velocity is 2860 ft. per second. Find the horizontal and vertical components of this velocity.

13. An airplane is flying a heading due west at an airspeed of 260 mi./hr. A wind is blowing directly from the north with a speed of 60 mi./hr. Find the drift angle, groundspeed, and course of the airplane.

13

THREE–DIMENSIONAL SPACE

127. Introduction. We live in a three-dimensional space and are accustomed to the fact that things have length, width, and height. Even without a formal course in solid geometry the reader is intuitively familiar with many of the properties of space. We shall begin our study of three-dimensional space from the point of view of analytic geometry.

128. Rectangular Cartesian Coordinates. The easiest way to visualize a coordinate system in space is to look into the corner of a room. Here we have three mutually perpendicular planes: the two walls and the floor. Thus we can locate a point as a certain distance from each of the walls and from the floor, so that it takes *three* real numbers which are usually denoted by x, y, and z. The three axes are called the X-axis, the Y-axis, and the Z-axis; the three coordinate planes are the XY-plane which contains the X- and Y-axes, the YZ-plane (the Y- and Z-axes), and the XZ-plane (the X- and Z-axes). The point P is located by the coordinates $P(x,y,z)$. The coordinate planes divide our three-dimensional space into eight parts called **octants.**

To picture the three dimensions on a plane figure we choose our axes as shown in Fig. 92, p. 202.

The **rules** *for signs*:

x is *positive* to the *front* of the YZ-plane; *negative* to the *back*.
y is *positive* to the *right* of the XZ-plane; *negative* to the *left*.
z is *positive above* the XY-plane; *negative below*.

The OY-axis is drawn perpendicular to the OZ-axis, and the OX-axis is usually drawn at a 45° angle from the YO-axis. The units on YY' and ZZ' are chosen to be equal, and the units on XX' are usually taken to one-half the units on YY'. On this kind of drawing, parallel lines will actually be drawn parallel. For illustration purposes the

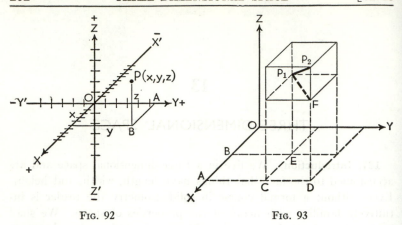

FIG. 92   FIG. 93

best picture is obtained in the first octant.

129. Distance between Two Points. We may obtain a formula for the distance between two points in space in the same manner as we did in a plane. Given any two points $P_1(x_1,y_1,z_1)$ and $P_2(x_2,y_2,z_2)$. Construct Fig. 93.

$$AB = x_2 - x_1, \quad CD = y_2 - y_1, \quad \text{and} \quad FP_2 = z_2 - z_1.$$

By the Pythagorean Theorem:

$$\begin{aligned}
\overline{P_1P_2}^2 &= \overline{P_1F}^2 + \overline{FP_2}^2 \\
&= \overline{EC}^2 + \overline{CD}^2 + \overline{FP_2}^2 \\
&= \overline{AB}^2 + \overline{CD}^2 + \overline{FP_2}^2 \\
&= (x_2 - x_1)^2 + (y_2 - y_1)^2 + (z_2 - z_1)^2.
\end{aligned}$$

The length of P_1P_2 is therefore given by:

(1) $$d = \sqrt{(x_2 - x_1)^2 + (y_2 - y_1)^2 + (z_2 - z_1)^2}.$$

From this formula it is easily seen that the distance from the origin $O(0,0,0)$ to a point $P(x,y,z)$ is given by:

(2) $$\rho = \sqrt{x^2 + y^2 + z^2}.$$

The directed segment OP whose length we just obtained is called the **radius vector** of P.

130. Direction Cosines of a Line. Consider the directed line OP, and let the angles formed by this line and the positive X-, Y-, and Z-axes be denoted by α, β, and γ. See Fig. 94.

These angles are called the *direction angles* of the line, and the cosines of these angles are called the **direction cosines** of the line. A close

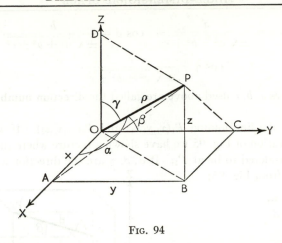

Fig. 94

inspection of the figure reveals that $\triangle OAP$ is a right triangle with
right angle at A; $\triangle OCP$ is a right triangle with right angle at C;
$\triangle ODP$ is a right triangle with right angle at D; and $OP = \rho$ is the
hypotenuse of each right triangle. Consequently, by definition we
have:

(1)　　　　$x = \rho \cos \alpha, \quad y = \rho \cos \beta, \quad z = \rho \cos \gamma.$

It is now a simple matter to prove that:
　　The sum of the squares of the direction cosines of any line equals one;
　　i.e.:

(2)　　　　　　$\cos^2 \alpha + \cos^2 \beta + \cos^2 \gamma = 1.$

The direction cosines of a line which does not pass through the origin
are defined as *those of the parallel line through the origin.*

131. Direction Numbers of a Line. In practice, three numbers
which are proportional to the direction cosines are frequently used.
Let a, b, c be three numbers such that:

$$\cos \alpha = ka, \quad \cos \beta = kb, \quad \cos \gamma = kc.$$

Then by formula (2) of Section 130 we have:

$$k^2a^2 + k^2b^2 + k^2c^2 = 1 = k^2(a^2 + b^2 + c^2)$$

so that:

$$k = \frac{1}{\pm \sqrt{a^2 + b^2 + c^2}}.$$

We then have:

$$\cos \alpha = \frac{a}{\pm \sqrt{a^2 + b^2 + c^2}}, \quad \cos \beta = \frac{b}{\pm \sqrt{a^2 + b^2 + c^2}},$$

$$\cos \gamma = \frac{c}{\pm \sqrt{a^2 + b^2 + c^2}}.$$

The numbers a, b, c used above are called the **direction numbers** of a line.

Consider the line joining $P_1(x_1, y_1, z_1)$ and $P_2(x_2, y_2, z_2)$. If we draw the configuration of Fig. 95 we have shown a picture where the origin may be considered to be at P_1. If α, β, γ are the direction angles of $P_1 P_2$, then from Fig. 95:

$\angle A P_1 P_2 = \alpha,$
$\angle B P_1 P_2 = \beta,$
$\angle C P_1 P_2 = \gamma,$

and

$$\cos \alpha = \frac{A P_1}{d} = \frac{x_2 - x_1}{d},$$

$$\cos \beta = \frac{B P_1}{d} = \frac{y_2 - y_1}{d},$$

$$\cos \gamma = \frac{C P_1}{d} = \frac{z_2 - z_1}{d}.$$

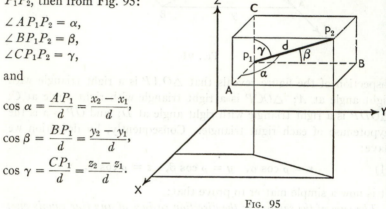

FIG. 95

Now since $d = \sqrt{(x_2 - x_1)^2 + (y_2 - y_1)^2 + (z_2 - z_1)^2}$ we have that the direction cosines of the line joining P_1 and P_2 are proportional to $x_2 - x_1$, $y_2 - y_1$, and $z_2 - z_1$, and these may be taken as the direction numbers of $P_1 P_2$.

132. Angle between Two Lines. The angle between two lines in space can be given in terms of the direction cosines. Consider the lines OP_1 and OP_2 shown in Fig. 96, with $\angle P_1 O P_2 = \theta$. By the law of cosines (see Section 120), we have:

$$\cos \theta = \frac{\rho_1^2 + \rho_2^2 - d^2}{2 \rho_1 \rho_2}.$$

Now:

$$\rho_1^2 = x_1^2 + y_1^2 + z_1^2,$$
$$\rho_2^2 = x_2^2 + y_2^2 + z_2^2,$$
$$d^2 = (x_2 - x_1)^2 + (y_2 - y_1)^2 + (z_2 - z_1)^2.$$
$$\therefore \quad \rho_1^2 + \rho_2^2 - d^2 = 2(x_1 x_2 + y_1 y_2 + z_1 z_2).$$

Furthermore by (1), Section 130:

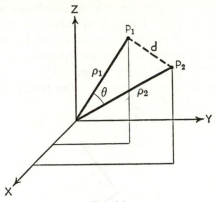

Fig. 96

$$x_1 = \rho_1 \cos \alpha_1, \quad x_2 = \rho_2 \cos \alpha_2, \quad y_1 = \rho_1 \cos \beta_1, \text{ etc.}$$

so that we have after simplifying:

(1)　　　$\cos \theta = \cos \alpha_1 \cos \alpha_2 + \cos \beta_1 \cos \beta_2 + \cos \gamma_1 \cos \gamma_2.$

If $\theta = 90°$, the two lines are perpendicular, and since $\cos 90° = 0$ we have: *Two lines in space are perpendicular if and only if the sum of the products of their corresponding direction cosines is zero;* i.e.:

(2)　　　$\cos \alpha_1 \cos \alpha_2 + \cos \beta_1 \cos \beta_2 + \cos \gamma_1 \cos \gamma_2 = 0.$

133. The Plane. In three-dimensional space the locus of an equation is a surface containing those points, and only those points, whose coordinates satisfy the equation. Thus if we relate the coordinates x, y, and z by an equation and give values to the coordinates, we find the points on a surface. The simplest surface is a plane. The reader should be familiar with the definition:

*A **plane** is a surface such that a straight line joining any two points of the surface lies entirely in the surface.*

We shall mainly be concerned with the analytical aspects of a plane. The fact that the equation:

$$x = c$$

represents a plane parallel to the YZ-plane at a distance c from it leads us to the statement:

An equation of the first degree in one variable represents a plane parallel to the plane of the other two variables when considered in three dimensions.

The converse is also true.

The intersections of a surface with the coordinate planes are called the **traces** of the surface. In order to obtain the traces in the XY-plane, we simply set $Z = 0$ in the given equation. Similarly for the other traces. The traces of a plane are straight lines.

Consider Fig. 97. ABC is any plane. $ON \perp$ to ABC at N with positive direction from O towards N. The length of $ON = p$. Let

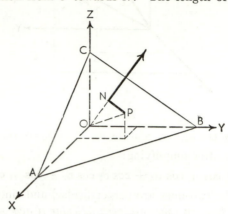

Fig. 97

α, β, γ be the direction angles of ON, and $P(x,y,z)$ be any point in the plane. The coordinates of N are given by:

$$x_1 = p \cos \alpha, \quad y_1 = p \cos \beta, \quad z_1 = p \cos \gamma.$$

The direction numbers of the line PN are:

$$x - p \cos \alpha, \quad y - p \cos \beta, \quad z - p \cos \gamma.$$

Now the lines ON and PN are perpendicular only if:

$$p \cos \alpha(x - p \cos \alpha) + p \cos \beta(y - p \cos \beta) + p \cos \gamma(z - p \cos \gamma) = 0$$

or

$$p(x \cos \alpha + y \cos \beta + z \cos \gamma) = p^2(\cos^2 \alpha + \cos^2 \beta + \cos^2 \gamma).$$

Referring back to the formulas of the last section again, we have:

(1) $$x \cos \alpha + y \cos \beta + z \cos \gamma = p.$$

This is the *normal form* of the equation of the plane.

We are now in a position to state the theorem:

Theorem. *The locus of the equation*:

(2) $$Ax + By + Cz + D = 0$$

is a plane.

Proof. Transpose the constant term D to the right member and make it positive. Now divide by $\sqrt{A^2 + B^2 + C^2} = k$. We have:

$$(3) \qquad \pm \frac{A}{k} x \pm \frac{B}{k} y \pm \frac{C}{k} z = \frac{\mp D}{k}.$$

Since

$$\left(\frac{A}{k}\right)^2 + \left(\frac{B}{k}\right)^2 + \left(\frac{C}{k}\right)^2 = \frac{A^2 + B^2 + C^2}{A^2 + B^2 + C^2} = 1$$

they satisfy the property of direction cosines of a line and consequently (3) is a normal form of the equation of a plane.

We are thus led to the rule:

Rule. *To reduce the equation of any plane:*

$$Ax + By + Cz + D = 0$$

to the normal form, divide by $\sqrt{A^2 + B^2 + C^2}$ and choose signs so that the constant is positive in the right member.

Since all equations of a plane can be put into normal form, it is easily seen that *two planes:*

$$A_1 x + B_1 y + C_1 z + D_1 = 0,$$
$$A_2 x + B_2 y + C_2 z + D_2 = 0$$

are perpendicular if:

$$A_1 A_2 + B_1 B_2 + C_1 C_2 = 0,$$

and conversely.

Illustration. Given the equation:

$$x + 2y + 3z = 6.$$

Find the intercepts on the axes, find the equations of the traces, draw the traces, and reduce the equation to normal form.

Solution.

Intercepts: $x = 6$, $y = 3$, $z = 2$.
XY-trace: $x + 2y = 6$.
YZ-trace: $2y + 3z = 6$.
XZ-trace: $x + 3z = 6$.
$A^2 + B^2 + C^2 = 1 + 4 + 9 = 14$.

Normal form:

$$\frac{1}{\sqrt{14}} x + \frac{2}{\sqrt{14}} y + \frac{3}{\sqrt{14}} z = \frac{6}{\sqrt{14}}.$$

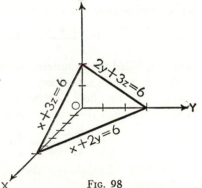

Fig. 98

The plane is shown in Fig. 98.

A plane is determined by three points.

Illustration. Find the equation of a plane through $(1,-3,2)$, $(3,1,4)$, and $(-1,-1,-2)$.

Solution. Substitute in the general form:

$$Ax + By + Cz + D = 0,$$

and let $D = 1$.

$$\begin{cases} A - 3B + 2C + 1 = 0. \\ 3A + B + 4C + 1 = 0. \\ -A - B - 2C + 1 = 0. \end{cases}$$

The solution of this system is:

$$A = -\tfrac{5}{2}, \quad B = \tfrac{1}{2}, \quad \text{and} \quad C = \tfrac{3}{2}.$$

The equation is:

$$-\tfrac{5}{2}x + \tfrac{1}{2}y + \tfrac{3}{2}z + 1 = 0$$

or

$$-5x + y + 3z + 2 = 0.$$

134. The Straight Line. In general, two surfaces intersect in a curve. If the surfaces are planes, the intersection is a straight line. Thus *the locus of two simultaneous equations of the first degree is a straight line,* and the **general form** of the equations of a straight line is:

$$(1) \qquad \begin{cases} A_1x + B_1y + C_1z + D_1 = 0, \\ A_2x + B_2y + C_2z + D_2 = 0. \end{cases}$$

It should be emphasized that it takes more than one equation to define a straight line in space. Another useful form of the equations of a straight line is the one obtained by considering a line through a point (x_1,y_1,z_1) with given direction numbers a, b, and c. Consider a general point (x,y,z) on the line; then the direction cosines are proportional to $x - x_1$, $y - y_1$, and $z - z_1$ as well as the direction numbers (see Section 131). Now since two sets of numbers proportional to the same third set are proportional to each other, we may write:

$$(2) \qquad \frac{x - x_1}{a} = \frac{y - y_1}{b} = \frac{z - z_1}{c},$$

which is called the **symmetric form** of the equations of a line.

We are now in a position to write the equations of a line through two points. Let there be given the points $P_1(x_1,y_1,z_1)$ and $P_2(x_2,y_2,z_2)$. The line through these points is given by the equations:

$$(3) \qquad \frac{x - x_1}{x_2 - x_1} = \frac{y - y_1}{y_2 - y_1} = \frac{z - z_1}{z_2 - z_1}.$$

Illustration. Find the equations of a line through $(3,-1,4)$ and $(1,3,-2)$.
Solution.

$$x_2 - x_1 = 1 - 3 = -2,$$
$$y_2 - y_1 = 3 + 1 = 4,$$
$$z_2 - z_1 = -2 - 4 = -6.$$

To reduce the equations of a line to a symmetric form eliminate any of the two variables, obtaining two equations which have the third variable in common; solve for this variable and equate the values; finally divide through by a number which will reduce the coefficient of each variable to unity.

Illustration. Reduce the equations:

$$x - 2y + 3z = -1, \quad 3x + 4y - 3z = 5$$

to the symmetric form.

Solution. Eliminate first y and then z.

$x - 2y + 3z = -1$	$2x - 4y + 6z = -2$
$3x + 4y - 3z = 5$	$3x + 4y - 3z = 5$
$4x + 2y = 4$	$5x + 3z = 3$
$x = \dfrac{-y + 2}{2}$	$x = \dfrac{-3z + 3}{5}$

$$\therefore \quad x = \frac{y - 2}{-2} = \frac{3(z - 1)}{-5}$$

$$\text{or} \quad \frac{x}{3} = \frac{y - 2}{-6} = \frac{z - 1}{-5}.$$

135. The Sphere. There are many surfaces in solid geometry. A special type is obtained by rotating a curve about a straight line. These are called **surfaces of revolution.** We shall simply give a table of the more common ones.

Surface	Generated by the rotation of
Sphere	A circle about a diameter.
Prolate Spheroid	An ellipse about its major axis.
Oblate Spheroid	An ellipse about its minor axis.
Paraboloid of revolution	A parabola about its axis.
Hyperboloid of revolution of one sheet	A hyperbola about its conjugate axis.
Hyperboloid of revolution of two sheets	A hyperbola about its transverse axis.
Circular cylinder	A straight line about a parallel line.
Circular cone	A straight line about a line intersecting it at an angle.

We shall limit our attention to a sphere.*

Definition. *A* **sphere** *is the locus of a moving point that remains at a constant distance, the* **radius,** *from a fixed point, the* **center.**

Let the center be at the point (h,k,l) and let the radius be r; then the equation of the sphere is:

$$(1) \qquad (x - h)^2 + (y - k)^2 + (z - l)^2 = r^2.$$

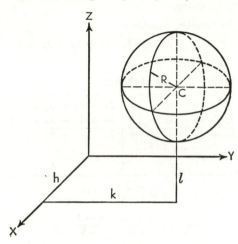

FIG. 99

This formula comes directly from the formula for the distance between two points.

Another form of the equation of a sphere is:

$$(2) \qquad x^2 + y^2 + z^2 + Gx + Hy + Iz + K = 0,$$

which can be reduced to equation (1) by completing the square in x, y, and z.

Illustration. Find the center and radius of the sphere:

$$x^2 + y^2 + z^2 - 2x + 6y - 4z - 2 = 0.$$

Solution. Complete the square in x, y, and z:

$$x^2 - 2x + 1 + y^2 + 6y + 9 + z^2 - 4z + 4 = 2 + 1 + 9 + 4.$$
$$(x - 1)^2 + (y + 3)^2 + (z - 2)^2 = 16.$$
$$\text{Center } (1,-3,2); \text{ radius} = 4.$$

---

* If the reader is interested in the other solids he is referred to C. O. Oakley, *Analytic Geometry* (New York: Barnes & Noble, Inc., 1957), pp. 198–223.

The sphere forms the fundamental model of an interesting subject in mathematics called *spherical trigonometry*. This subject finds applications in astronomy, navigation, military aspects of aerial combat, and studies in physical systems. The subject matter of spherical trigonometry is not beyond the level of this book, but to study it properly more time and space would be needed than can be permitted here.*

136. Exercise XIII.

1. Plot the points $P_1(8,3,4)$, $P_2(5,-3,2)$, and $P_3(-4,-7,-3)$.
2. Find the distance between each of the points of Problem 1.
3. Find the direction numbers of the line between $P_1(8,3,4)$ and $P_2(5,-3,2)$ and write the equations of this line.
4. Given the equations:

$$x - 2y + 3z = 6,$$
$$x + 2y + z = 4.$$

For each plane find the intercepts on the axes, find the equations of the traces, and reduce to normal form. Prove that the planes are perpendicular, and reduce the equations of the line of intersection to the symmetric form.

5. Find the center and radius of the sphere:

$$2x^2 + 2y^2 + 2z^2 - 4x + 5y - 6z = 0.$$

---

* See K. L. Nielsen and J. H. VanLonkhuyzen, *Plane and Spherical Trigonometry* (New York: Barnes & Noble, Inc., 1954), Part Two.

14

THE CALCULUS

137. Introduction. A student is usually considered to have reached a certain mathematical maturity after he has completed a course in the calculus. This mathematical maturity is the minimum requirement for an engineer. The study of the calculus is quite fascinating, for here we put to application the finer theories of mathematics and explain many of the laws of nature through the language of mathematics. It is, of course, beyond the scope of this book to consider the entire course in the calculus. However, we shall present an insight and hope to stimulate the reader to pursue the study properly.*

138. Limits. In Chapter 2 we defined a function, variables and constants, and the concept of an increment. Let us recall that if:

$$y = f(x)$$

and we let the independent variable x change by an increment Δx, then Δy will denote the corresponding increment of the function $f(x)$.

Consider the function:

(1) $$y = x^2 + x.$$

Let us start with a given initial value of x, say x_0, and let x be changed by an increment Δx; then:

$$
\begin{aligned}
y + \Delta y &= (x + \Delta x)^2 + (x + \Delta x) \\
&= x^2 + 2x(\Delta x) + (\Delta x)^2 + x + \Delta x
\end{aligned}
$$

Subtract (1): $\quad\underline{\quad y = x^2 \qquad\qquad\qquad + x\quad}$

$$\Delta y = 2x(\Delta x) + (\Delta x)^2 + \Delta x$$

Divide by Δx:

$$\frac{\Delta y}{\Delta x} = 2x + 1 + \Delta x.$$

Let us investigate the behavior of this ratio as we change the incre-

---

* See C. O. Oakley, *The Calculus* (New York: Barnes & Noble, Inc., 1957), pp. 24–28.

ment Δx so that x gets closer and closer to our picked initial value; i.e., as Δx approaches zero.

x_0	x	Δx	$\dfrac{\Delta y}{\Delta x}$
2	3	1	8
2	2.5	.5	6.5
2	2.1	.1	5.3
2	2.05	.05	5.15
2	2.01	.01	5.03

We note that the ratio $\Delta y/\Delta x$ is approaching 5. If we can bring the ratio as close to 5 as we please by taking Δx smaller and smaller, we say that the limit of the ratio as Δx approaches zero is 5; symbolically:

$$\lim_{\Delta x \to 0} \frac{\Delta y}{\Delta x} = 5.$$

This is a very important concept and is fundamental in the study of the calculus. The rigorous definition states:

A constant A is said to be the limit of a variable Z if, as the variable changes its value according to some formula, the numerical difference between the variable and the constant, $|Z - A|$, becomes and remains less than any small positive quantity, $\epsilon > 0$, which may be assigned. See also Section 92.

Illustration. Let us find the value of:

$$1 + \tfrac{1}{2} + \tfrac{1}{4} + \tfrac{1}{8} + \cdots$$

as we take an infinite number of terms.

Solution. This is the sum of an unlimited geometric progression (Section 92) and:

$$S = \frac{a}{1 - r} = \frac{1}{1 - \tfrac{1}{2}} = 2.$$

Thus the limit of S is 2.

We shall now state without proof three theorems on limits. In the following let u, v, and w be functions of x, and let:

$$\lim_{x \to a} u = A, \quad \lim_{x \to a} v = B, \quad \lim_{x \to a} w = C.$$

Then:

(1) $\lim\limits_{x \to a} (u + v - w) = A + B - C,$

(2) $\lim\limits_{x \to a} (uvw) = ABC,$

(3) $\lim\limits_{x\to a} \dfrac{u}{v} = \dfrac{A}{B}$, if $B \neq 0$.

Furthermore, if c is a constant, $c \neq 0$:

$\lim\limits_{v\to 0} \dfrac{c}{v} = \infty$	$\lim\limits_{v\to\infty} cv = \infty$
$\lim\limits_{v\to\infty} \dfrac{v}{c} = \infty$	$\lim\limits_{v\to\infty} \dfrac{c}{v} = 0$

139. The Derivative. Let us consider a function of one variable:

(1) $$y = f(x),$$

and let us form the quotient of the two increments:

$$\begin{aligned} y + \Delta y &= f(x + \Delta x) \\ (-) \quad y \quad\;\; &= f(x) \\ \hline \Delta y &= f(x + \Delta x) - f(x) \end{aligned}$$

and

$$\frac{\Delta y}{\Delta x} = \frac{f(x + \Delta x) - f(x)}{\Delta x}.$$

Now consider the limit of this ratio as Δx approaches zero. This leads us to the fundamental definition of the differential calculus:

The **derivative** *of a function is the limit of the ratio of the increment of the function, Δy, to the increment of the independent variable, Δx, when Δx varies and approaches zero as a limit.*

The symbol:

$$\frac{dy}{dx} = \lim_{\Delta x\to 0} \frac{f(x + \Delta x) - f(x)}{\Delta x}$$

defines the *derivative of y with respect to x.* It is not a fraction but is the limiting value of a fraction. Thus the symbol dy/dx is to be considered as a whole. There are other symbols for the derivative, and one which is prominent is:

$$\frac{dy}{dx} = f'(x);$$

i.e., "the derivative of y with respect to x is f prime of x."

The symbol:

$$\frac{d}{dx}(\;)$$

is called the **differential operator**. Another symbol is D_x. Thus we have the following identical symbols:

$$y' = \frac{dy}{dx} = \frac{d}{dx}(y) = \frac{d}{dx}[f(x)] = D_x f(x) = f'(x).$$

When the limit of the ratio $\Delta y / \Delta x$ exists, the function is said to be *differentiable* or to *possess a derivative*. In order for this to be true the function must be continuous and, in general, we shall be concerned only with functions which are differentiable.

The Four-Step Rule.

First Step. In the function, $f(x)$, replace x by $x + \Delta x$ and calculate $y + \Delta y$.

Second Step. Subtract the given value of the function from the new value and find Δy.

Third Step. Divide Δy by Δx.

Fourth Step. Find the limit of this quotient when Δx varies and approaches zero as a limit.

Illustration. Differentiate $2x^2 + 3x + 2$.
Solution.
$$y = 2x^2 + 3x + 2.$$

1. $\quad y + \Delta y = 2(x + \Delta x)^2 + 3(x + \Delta x) + 2$
$$= 2x^2 + 4x(\Delta x) + 2(\Delta x)^2 + 3x + 3\Delta x + 2$$

2. $\quad \underline{y = 2x^2 \qquad\qquad\qquad + 3x \qquad\qquad + 2}$
$$\Delta y = 4x(\Delta x) + 2(\Delta x)^2 \qquad\qquad + 3\Delta x$$

3. $\quad \dfrac{\Delta y}{\Delta x} = 4x + 2\Delta x + 3.$

4. $\quad \dfrac{dy}{dx} = 4x + 3.$

Although the four-step rule will obtain any derivative, it becomes too cumbersome to perform each time, and we shall derive some formulas for application.

140. Derivative Formulas. We shall apply the general rule to some standard functional forms and obtain the necessary formulas. The student should commit these formulas to memory. In the notation c is a constant and u, v, and w are differentiable functions of x.

Formulas for the Algebraic Forms.

I. $\qquad \dfrac{dc}{dx} = 0.$

II. $\dfrac{dx}{dx} = 1.$

III. $\dfrac{d}{dx}(u + v - w) = \dfrac{du}{dx} + \dfrac{dv}{dx} - \dfrac{dw}{dx}.$

IV. $\dfrac{d}{dx}(cv) = c\,\dfrac{dv}{dx}.$

V. $\dfrac{d}{dx}(uv) = u\,\dfrac{dv}{dx} + v\,\dfrac{du}{dx}.$

VI. $\dfrac{d}{dx}(v^n) = nv^{n-1}\,\dfrac{dv}{dx}.$

VIa. $\dfrac{d}{dx}(x^n) = nx^{n-1}.$

VII. $\dfrac{d}{dx}\left(\dfrac{u}{v}\right) = \dfrac{v\,\dfrac{du}{dx} - u\,\dfrac{dv}{dx}}{v^2}.$

VIIa. $\dfrac{d}{dx}\left(\dfrac{u}{c}\right) = \dfrac{1}{c}\,\dfrac{du}{dx}.$

VIII. $\dfrac{dy}{dx} = \dfrac{dy}{dv} \cdot \dfrac{dv}{dx},$ where $y = f(v).$

IX. $\dfrac{dy}{dx} = \dfrac{1}{\dfrac{dx}{dy}},$ where $y = f(x).$

These formulas may be derived by applying the four-step rule.

I. $y = c.$
$y + \Delta y = c.$
$\Delta y = 0.$
$\lim\limits_{\Delta x \to 0} \dfrac{\Delta y}{\Delta x} = \dfrac{dy}{dx} = \dfrac{dc}{dx} = 0.$

The derivative of a constant is zero.

III. $y = u + v - w.$
$y + \Delta y = u + \Delta u + v + \Delta v - w - \Delta w.$
$\Delta y = \Delta u + \Delta v - \Delta w.$
$\dfrac{\Delta y}{\Delta x} = \dfrac{\Delta u}{\Delta x} + \dfrac{\Delta v}{\Delta x} - \dfrac{\Delta w}{\Delta x}.$
$\lim\limits_{\Delta x \to 0} \dfrac{\Delta y}{\Delta x} = \lim\limits_{\Delta x \to 0}\left(\dfrac{\Delta u}{\Delta x} + \dfrac{\Delta v}{\Delta x} - \dfrac{\Delta w}{\Delta x}\right).$
$\dfrac{dy}{dx} = \dfrac{du}{dx} + \dfrac{dv}{dx} - \dfrac{dw}{dx},$

by Sections 138 and 139.

The derivative of the algebraic sum of n functions is equal to the same algebraic sum of their derivatives.

Formulas I through IX may all be obtained in this manner. It is most advantageous to remember these formulas in terms of words as well as symbols. For example, formula VII states:

The derivative of a quotient is equal to the denominator times the derivative of the numerator, minus the numerator times the derivative of the denominator, all divided by the square of the denominator.

It is left to the student to express each of the formulas in terms of words, and it is urged that he do so to make sure he understands the concepts.

Formulas for Transcendental Functions.

X. $\dfrac{d}{dx}(\log_e v) = \dfrac{1}{v}\dfrac{dv}{dx}.$

Xa. $\dfrac{d}{dx}(\log_{10} v) = \dfrac{\log_{10} e}{v}\dfrac{dv}{dx}.$

XI. $\dfrac{d}{dx}(a^v) = a^v \log_e a \dfrac{dv}{dx}.$

XIa. $\dfrac{d}{dx}(e^v) = e^v \dfrac{dv}{dx}.$

XII. $\dfrac{d}{dx}(u^v) = vu^{v-1}\dfrac{du}{dx} + \log_e u \; u^v \dfrac{dv}{dx}.$

XIII. $\dfrac{d}{dx}(\sin v) = \cos v \dfrac{dv}{dx}.$

XIV. $\dfrac{d}{dx}(\cos v) = -\sin v \dfrac{dv}{dx}.$

XV. $\dfrac{d}{dx}(\tan v) = \sec^2 v \dfrac{dv}{dx}.$

XVI. $\dfrac{d}{dx}(\cot v) = -\csc^2 v \dfrac{dv}{dx}.$

XVII. $\dfrac{d}{dx}(\sec v) = \sec v \tan v \dfrac{dv}{dx}.$

XVIII. $\dfrac{d}{dx}(\csc v) = -\csc v \cot v \dfrac{dv}{dx}.$

XIX. $\dfrac{d}{dx}(\arcsin v) = \dfrac{1}{\sqrt{1-v^2}}\left(\dfrac{dv}{dx}\right).$

XX. $\dfrac{d}{dx}(\text{arc cos } v) = -\dfrac{1}{\sqrt{1-v^2}}\left(\dfrac{dv}{dx}\right).$

XXI. $\dfrac{d}{dx}(\text{arc tan } v) = \dfrac{1}{1+v^2}\left(\dfrac{dv}{dx}\right).$

XXII. $\dfrac{d}{dx}(\text{arc cot } v) = -\dfrac{1}{1+v^2}\left(\dfrac{dv}{dx}\right).$

XXIII. $\dfrac{d}{dx}(\text{arc sec } v) = \dfrac{1}{v\sqrt{v^2-1}}\left(\dfrac{dv}{dx}\right).$

XXIV. $\dfrac{d}{dx}(\text{arc csc } v) = -\dfrac{1}{v\sqrt{v^2-1}}\left(\dfrac{dv}{dx}\right).$

The derivation of these formulas is often made to depend upon two limits; namely:

$$\lim_{v\to 0}\frac{\sin v}{v} = 1$$

and

$$\lim_{v\to 0}(1+v)^{\frac{1}{v}} = e.$$

If we know these two limits the above formulas may be derived by application of the known formulas and the four-step rule. We shall simply assume the truth of the formulas and apply them to some problems.*

Illustrations. Find the derivative $\dfrac{dy}{dx}$ of:

1. $y = 12x^3 - 6x + 2.$

Solution.

$$\frac{dy}{dx} = \frac{d}{dx}(12x^3) - \frac{d}{dx}(6x) + \frac{d}{dx}(2)$$

$$= 12\left[3x^2\frac{dx}{dx}\right] - 6\left(\frac{dx}{dx}\right) + 0$$

$$= 36x^2 - 6.$$

2. $y = (x^2 + 3)\sqrt{2x - 1}.$

Solution.

$$y' = (x^2 + 3)\frac{d}{dx}\left(\sqrt{2x-1}\right) + \sqrt{2x-1}\left(\frac{d}{dx}[x^2 + 3]\right)$$

$$= (x^2 + 3)\tfrac{1}{2}(2x-1)^{-\frac{1}{2}}(2) + \sqrt{2x-1}(2x)$$

$$= \frac{x^2 + 3}{\sqrt{2x-1}} + 2x\sqrt{2x-1}$$

---

* See any book on *The Calculus* for complete derivations of each formula.

$$y' = \frac{x^2 + 3 + 2x(2x - 1)}{\sqrt{2x - 1}}$$

$$= \frac{5x^2 - 2x + 3}{\sqrt{2x - 1}}.$$

3. $x^2 - \sqrt{xy} + y = a$.

Solution.

$$2x \frac{dx}{dx} - \frac{1}{2} (xy)^{-\frac{1}{2}} \left[y \frac{dx}{dx} + x \frac{dy}{dx} \right] + \frac{dy}{dx} = 0.$$

$$2x - \frac{1}{2}(xy)^{-\frac{1}{2}}y - \frac{1}{2}(xy)^{-\frac{1}{2}}xy' + y' = 0.$$

$$\left(1 - \frac{x}{2\sqrt{xy}} \right) y' = \frac{y}{2\sqrt{xy}} - 2x.$$

$$\frac{dy}{dx} = y' = \frac{y - 4x\sqrt{xy}}{2\sqrt{xy} - x}.$$

4. $y = 4 \sin ax^3$.

Solution.

$$\frac{dy}{dx} = 4 \cos ax^3 \left[\frac{d}{dx} (ax^3) \right]$$

$$= 4(3ax^2) \cos ax^3$$

$$= 12ax^2 \cos ax^3.$$

5. $y = e^{-x} \cos x$.

Solution.

$$\frac{dy}{dx} = e^{-x} \left[\frac{d}{dx} (\cos x) \right] + \cos x \left[\frac{d}{dx} (e^{-x}) \right]$$

$$= e^{-x}(-\sin x) + \cos x(e^{-x})(-1)$$

$$= -e^{-x} [\sin x + \cos x].$$

6. $y = \arctan \sqrt{x}$.

Solution.

$$y' = \frac{1}{1 + x} \left[\frac{d}{dx} (\sqrt{x}) \right]$$

$$= \frac{1}{1 + x} (\tfrac{1}{2}x^{-\frac{1}{2}})$$

$$= \frac{1}{2\sqrt{x}(1 + x)}.$$

To familiarize himself with the formulas the student should work more than 100 exercises. It is easy to make up these exercises by writing down any function that may occur to one and then finding the derivative with respect to the chosen independent variable.

141. Successive Differentiation. The derivative of a function of x may also be a function of x, and thus be differentiable. If we differentiate the *first derivative*, we obtain the **second derivative** of the original function. The derivative of the *second derivative* is the **third derivative**; etc. These are called **higher derivatives,** and we use the notation:

$$\frac{d}{dx}\left(\frac{dy}{dx}\right) = \frac{d^2y}{dx^2}; \quad \frac{d}{dx}\left(\frac{d^2y}{dx^2}\right) = \frac{d^3y}{dx^3}; \quad \text{etc.}$$

or

$$\frac{dy}{dx} = y' = f'(x); \quad \frac{d^2y}{dx^2} = y'' = f''(x); \quad \text{etc.}$$

In general we have:

$$\frac{d^ny}{dx^n} = y^{(n)} = f^{(n)}(x).$$

Illustration 1. Find the third derivative of:

$$y = 6x^5 + 3x^3 + 6x^2.$$

Solution.

$$y' = 30x^4 + 9x^2 + 12x,$$
$$y'' = 120x^3 + 18x + 12,$$
$$y''' = 360x^2 + 18.$$

Illustration 2. Find $\frac{d^2y}{dx^2}$ for the equation of the hyperbola, $b^2x^2 - a^2y^2 = a^2b^2$.

Solution. Differentiate with respect to x:

$$2b^2x - 2a^2yy' = 0$$

or

$$y' = \frac{b^2x}{a^2y}.$$

Differentiate again:

$$\frac{d^2y}{dx^2} = \frac{a^2y(b^2) - b^2x(a^2y')}{a^4y^2}$$

$$= \frac{a^2b^2y - a^2b^2x\left(\dfrac{b^2x}{a^2y}\right)}{a^4y^2}$$

$$= -\frac{b^2(b^2x^2 - a^2y^2)}{a^4y^3}.$$

However, $b^2x^2 - a^2y^2 = a^2b^2$, so that:

$$\frac{d^2y}{dx^2} = -\frac{b^4}{a^2y^3}.$$

Illustration 3. Find y''' for $y = e^x$.

Solution.

$$y = e^x,$$
$$y' = e^x(1) = e^x,$$

$$y'' = e^x(1) = e^x,$$
$$y''' = e^x(1) = e^x.$$

It is interesting to note that the derivative of $y = e^x$ is the same function over again. The exponential function is unique in this respect.

142. Applications of the Derivative. The concept of differentiation has many applications, and it is possible to consider only a few of them here.

A. Slope of the tangent line. The derivative of the function of a curve, $y = f(x)$, is the slope of the tangent line to the curve at $P(x,y)$. If we let α be the inclination of the tangent line, then:

$$\frac{dy}{dx} = \tan \alpha = \textbf{ slope of the curve at any point } P(x,y).$$

If $\dfrac{dy}{dx} = 0$, then $\tan \alpha = 0$ or $\alpha = 0$, and the tangent line is *horizontal*. If $\dfrac{dy}{dx}$ becomes infinite, then $\alpha = 90°$, and the tangent line is *vertical*.

B. Maximum and minimum values of a function. A maximum value of a function is one that is *greater* than any value immediately preceding or following. A **minimum** value of a function is one that is *less* than any value immediately preceding or following.

To find the maximum and minimum values of a function, find the first derivative and set it equal to zero. The solutions of this equation yield the critical values of the function. Let x_1 be such a critical value; then:

$f(x)$ **is a maximum if** $f'(x_1 - \epsilon) > 0$ **and** $f'(x_1 + \epsilon) < 0,$
$f(x)$ **is a minimum if** $f'(x_1 - \epsilon) < 0$ **and** $f'(x_1 + \epsilon) > 0,$

where $f'(x_1 - \epsilon)$ is the value of the derivative for a value of x just less than x_1 and $f'(x_1 + \epsilon)$ is the value of the derivative for a value of x just more than x_1.

Illustration. Find the maximum and minimum values of:

$$y = 4x^3 + 9x^2 - 12x + 5.$$

Solution. Set the first derivative equal to zero:
$$y' = f'(x) = 12x^2 + 18x - 12 = 0.$$

The solutions are:
$$x_1 = \tfrac{1}{2} \quad \text{and} \quad x_2 = -2.$$

Now: $$f'(0) = -12, f'(1) = +18.$$

∴ x_1 yields a minimum.
$$f'(-3) = 42, f'(0) = -12.$$
∴ x_2 yields a maximum.
The minimum value is $f(\frac{1}{2}) = \frac{7}{4}$.
The maximum value is $f(-2) = 33$.

C. Point of inflection. It is possible to think of a curve as being made up of a series of arcs. That point on the curve which separates arcs of opposite directions of bending is called a **point of inflection.** It is also the point at which the tangent crosses the curve. To find the points of inflection we rely on the property:

At points of inflection, $f''(x) = 0$.

However, we must be careful, as this is a point of inflection only if the second derivative changes sign as we pass through the point. Thus to find the point of inflection:

1. Find the second derivative.
2. Set $f''(x) = 0$ and find the real roots.
3. Test $f''(x)$ for values in the neighborhood of these roots; if it changes sign we have a point of inflection.
4. If $f''(x) > 0$, the curve is concaved *upward*, and if $f''(x) < 0$, the curve is concaved *downward*.

Illustration. Find the point of inflection of $y = x^3$.
Solution. Set:
$$y'' = 0.$$
$$y'' = 6x = 0.$$

The critical value is at $x = 0$. Now:
$$y''(-1) = 6(-1) = -6$$
$$y''(1) = 6(1) = 6.$$

∴ curve is concaved downward to the left of $x = 0$ and concaved upward to the right of $x = 0$, and $(0,0)$ is a point of inflection.

D. Curve tracing. We are now in a good position to trace a curve; i.e., plot it on a coordinate system. The procedure is as follows (we shall limit ourselves to rectangular coordinates):

1. Find the first derivative, set it equal to zero, and solve for critical values of the abscissa.

2. Find the second derivative, set it equal to zero, and solve for critical values of the abscissa.

3. Surround the critical values with convenient values of the abscissa and calculate the values of the function, and the first and second derivatives.

4. Make a table of the values (see illustration) and plot the points

Illustration. Plot the graph of:

$$y = x^3 - 3x^2 - 45x + 108.$$

Solution.

$$y' = 3x^2 - 6x - 45.$$
$$y' = 0 \text{ yields } x = -3, x = 5.$$
$$y'' = 6x - 6.$$
$$y'' = 0 \text{ yields } x = 1.$$

x	y	y'	y''	Comments
-4	176	$+$	$-$	
-3	189	0	$-$	Max. } Concaved down
-1	149	$-$	$-$	
0	108	$-$	$-$	
1	61	$-$	0	Point of inflection
3	-27	$-$	$+$	
5	-67	0	$+$	Min. } Concaved up
6	-54	$+$	$+$	

The graph is shown in Fig. 100.

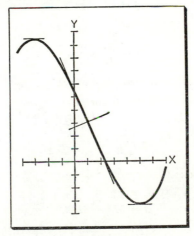

Fig. 100

E. Velocity and acceleration. Let us consider rectilinear motion, i.e., motion in a straight line. The distance traveled is the time taken multiplied by the rate, or we say, the distance is a function of time; thus:

$$s = f(t).$$

If we let:

$$\frac{\Delta s}{\Delta t} = \text{the average velocity,}$$

we see that the instantaneous velocity is the limit of the average velocity as $\Delta t \to 0$.

Thus:

$$v = \frac{ds}{dt}.$$

Similarly, we define *acceleration* as *the time-rate of change of the velocity*, and:

$$\text{Acceleration} = a = \frac{dv}{dt} = \frac{d^2s}{dt^2}.$$

Illustration. If we neglect the aerodynamic forces, the distance reached by a body projected vertically upwards with a velocity of v_1 ft. per sec. is given by $s = v_1t - \frac{1}{2}gt^2$, where $g = 32.2$. Find the maximum height reached, the acceleration, and the velocity at the end of 3 seconds if $v_1 = 322$ ft./sec.

Solution. Since:

$$s = v_1t - \tfrac{1}{2}gt^2,$$
$$v = \frac{ds}{dt} = v_1 - gt,$$
$$a = \frac{d^2s}{dt^2} = -g,$$

the maximum height is reached when $v = 0$.

$$v_1 - gt = 0 \quad \text{or} \quad t = v_1/g.$$
$$s_{\max} = v_1\left(\frac{v_1}{g}\right) - \frac{1}{2}g\left(\frac{v_1}{g}\right)^2$$
$$= \frac{1}{g}\left(v_1^2 - \frac{1}{2}v_1^2\right) = \frac{1}{2g}v_1^2.$$

At $v_1 = 322$ ft./sec.:

$$a = -g = -32.2.$$
$$v = v_1 - gt = 322 - 32.2t.$$
$$s_{\max} = \frac{1}{2g}(322)^2 = \frac{1}{2(32.2)}(322)^2 = 1610 \text{ ft.}$$

At $t = 3$ sec.:

$$v = 322 - 32.2(3) = 225.4 \text{ ft./sec.}$$

The concepts of velocity and acceleration have many applications in our scientific world. The flight of airplanes, the motion of machinery, the flow of electricity, etc. are all examples. The theory of derivatives and integrals (which we shall soon introduce) permits the solution of many complicated problems. If the student has an interest along these lines a thorough study of the calculus is recommended.

143. The Integral. There are many operations in mathematics which are spoken of as *inverse* operations, for example, addition and subtraction, multiplication and division, exponents and logarithms. We shall now discuss another inverse relation. Let us suppose we have given the derivative of a function and we desire to find the function itself. The function, $f(x)$, which we find is called the **integral** of the given differential expression, and the process is called **integration.** Symbolically:

$$\int f'(x)\, dx = f(x)$$

reads *the integral of $f'(x)$ dx equals $f(x)$.* The sign $\int$ is called the **integral sign,** and the expression dx is called a **differential** and indicates that x is the *variable of integration.*

Consider the function $f(x) = x^4$. The derivative is $f'(x) = 4x^3$ and we have:

$$\int f'(x)\, dx = \int 4x^3\, dx = x^4.$$

Consider now the function $f(x) = x^4 + 5$, whose derivative is

$$f'(x) = 4x^3,$$

and we have:

$$\int f'(x)\, dx = \int 4x^3\, dx = x^4 + 5.$$

It appears that there is more than one function which can be obtained by integration. However, it may be noted that the difference between the two answers is a constant, 5, and we recall that the derivative of a constant is zero. It is thus easily seen that two functions which have the same derivative can differ by an indefinite constant. This is expressed by the statement:

$$\int f'(x)\, dx = f(x) + c$$

and is called the **indefinite integral of $f'(x)$ dx.**

144. Elementary Integral Forms. There is no set rule for finding the integral of a given differential. We are in each case seeking *the function which, when differentiated, will yield the given differential expression.* To do this we rely on our knowledge of the derivative and have developed a set of integral formulas for the expression which can easily be obtained.*

---

* Extensive tables of integrals have been published. See R. S. Burington, *Handbook of Mathematical Tables and Formulas* (Sandusky, Ohio: Handbook Publishers, Inc., 1943).

Elementary Forms of the Indefinite Integral *

(1) $\int (du + dv - dw) = \int du + \int dv - \int dw.$

(2) $\int a\, du = a \int du.$

(3) $\int dx = x + c.$

(4) $\int u^n\, du = \dfrac{u^{n+1}}{n+1} + c, \quad (n \neq -1).$

(5) $\int \dfrac{du}{u} = \log_e u + c.$

(6) $\int a^u\, du = \dfrac{a^u}{\log_e a} + c.$

(7) $\int e^u\, du = e^u + c.$

(8) $\int \sin u\, du = -\cos u + c.$

(9) $\int \cos u\, du = \sin u + c.$

(10) $\int \sec^2 u\, du = \tan u + c.$

(11) $\int \csc^2 u\, du = -\cot u + c.$

(12) $\int \sec u \tan u\, du = \sec u + c.$

(13) $\int \csc u \cot u\, du = -\csc u + c.$

(14) $\int \tan u\, du = \log_e \sec u + c.$

(15) $\int \cot u\, du = \log_e \sin u + c.$

(16) $\int \sec u\, du = \log_e (\sec u + \tan u) + c.$

(17) $\int \csc u\, du = \log_e (\csc u - \cot u) + c.$

(18) $\int \dfrac{du}{u^2 + a^2} = \dfrac{1}{a} \arctan \dfrac{u}{a} + c.$

(19) $\int \dfrac{du}{u^2 - a^2} = \dfrac{1}{2a} \log_e \dfrac{u-a}{u+a} + c.$

(20) $\int \dfrac{du}{a^2 - u^2} = \dfrac{1}{2a} \log_e \dfrac{a+u}{a-u} + c.$

(21) $\int \dfrac{du}{\sqrt{a^2 - u^2}} = \arcsin \dfrac{u}{a} + c.$

(22) $\int \sqrt{a^2 - u^2}\, du = \dfrac{u}{2}\sqrt{a^2 - u^2} + \dfrac{a^2}{2} \arcsin \dfrac{u}{a} + c.$

The truth of these formulas can be checked by differentiating the right-hand expressions. Let us apply these formulas to some examples.

Illustrations. Perform the indicated integration.

1. $\int 21x^6\, dx.$

* In all the formulas involving log x we shall assume $x > 0$.

Solution. By formulas (2) and (4):

$$\int 21x^6 \, dx = 21 \int x^6 \, dx = 21 \frac{x^7}{7} + c = 3x^7 + c.$$

2. $\int \left(\frac{x^3}{3} - \frac{3}{x^3} \right) dx.$

Solution. By formulas (1), (2), and (4):

$$\int \left(\frac{x^3}{3} - \frac{3}{x^3} \right) dx = \int \frac{x^3}{3} \, dx - \int 3x^{-3} \, dx$$

$$= \frac{x^4}{12} - 3 \frac{x^{-2}}{-2} + c$$

$$= \frac{x^4}{12} + \frac{3}{2x^2} + c.$$

3. $\int \frac{\sec^2 x \, dx}{2 + 3 \tan x}.$

Solution. Differentiate the denominator:

$$\frac{d}{dx} (2 + 3 \tan x) = 3 \sec^2 x.$$

Thus the numerator is the differential of the denominator if we multiply it by 3. Let us therefore multiply both numerator and denominator by 3 and apply formulas (2) and (5) with $u = 2 + 3 \tan x$.

$$\int \frac{\sec^2 x \, dx}{2 + 3 \tan x} = \int \frac{3 \sec^2 x \, dx}{3(2 + 3 \tan x)}$$

$$= \frac{1}{3} \int \frac{3 \sec^2 x \, dx}{2 + 3 \tan x}$$

$$= \tfrac{1}{3} \log_e (2 + 3 \tan x) + c.$$

4. $\int \frac{x^2 - 6}{x - 3} \, dx.$

Solution. First divide out the fraction.

$$\frac{x^2 - 6}{x - 3} = x + 3 + \frac{3}{x - 3}$$

Apply formulas (2), (4), (3), and (5):

$$\int \frac{x^2 - 6}{x - 3} \, dx = \int x \, dx + 3 \int dx + 3 \int \frac{dx}{x - 3}$$

$$= \frac{x^2}{2} + 3x + 3 \log_e (x - 3) + c.$$

To apply the integration formulas it is very important to be able to pick out u and the corresponding du. Frequently part of the function expressed belongs to du as in Illustration 3 above. Another example will be shown.

Illustration. Find $\int (3 - 2x^2)^{\frac{3}{2}} x\, dx$.

Solution. Let:

$$u = 3 - 2x^2;$$

then:

$$\frac{du}{dx} = -4x$$

and:

$$du = -4x\, dx.$$

If we multiply and divide by -4, we may write:

$$\int (3 - 2x^2)^{\frac{3}{2}} x\, dx = -\tfrac{1}{4} \int (3 - 2x^2)^{\frac{3}{2}} (-4x\, dx)$$

which is of the form given by formula (4). Thus:

$$\int (3 - 2x^2) x\, dx = -\frac{1}{4} \frac{(3 - 2x^2)^{\frac{5}{2}}}{\frac{5}{2}} + c$$

$$= -\tfrac{1}{10}(3 - 2x^2)^{\frac{5}{2}} + c.$$

145. The Definite Integral. We shall now introduce a new concept:

$$(1) \qquad \int_a^b f'(x)\, dx = f(b) - f(a).$$

The symbol:

$$\int_a^b f'(x)\, dx$$

is read "the integral from a to b of $f'(x)\, dx$" and is called a **definite integral.** It defines an operation called *integration between limits.* The letter written at the bottom, a, is called the **lower limit,** and the one written at the top, b, is called the **upper limit.** It is clear from equation (1) that the definite integral always has a definite value and that the constant of integration has disappeared. To find the value of $\int_a^b f'(x)\, dx$, first evaluate $\int f'(x)\, dx$, then substitute $x = b$ and $x = a$ into the resulting function, and finally subtract the latter result from the first.

Illustration. Evaluate $\int_2^4 (x^2 + x)\, dx$.

Solution.

$$\int_2^4 (x^2 + x)\, dx = \left[\frac{x^3}{3} + \frac{x^2}{2} \right]_2^4$$

$$= \left[\frac{4^3}{3} + \frac{4^2}{2} \right] - \left[\frac{2^3}{3} + \frac{2^2}{2} \right]$$

$$= 16[\tfrac{4}{3} + \tfrac{1}{2}] - 4[\tfrac{2}{3} + \tfrac{1}{2}] = \tfrac{74}{3}.$$

146. Application of the Integral. There are numerous applications of the integral. We shall consider only one. Let:

$$y = f(x)$$

be the equation of a curve which is continuous and single-valued. Suppose we can draw it as shown in Fig. 101.

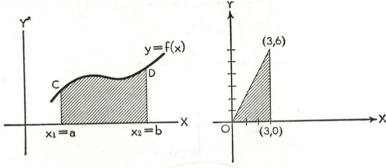

FIG. 101   FIG. 102

Choose two values of x at $x_1 = a$ and $x_2 = b$ and draw the vertical lines. We now have an area bounded by the curve, the X-axis, and these vertical lines; i.e., the area $x_1 CD x_2$. It can be shown that:

$$\textbf{Area } x_1 CD x_2 = F(b) - F(a) = \int_a^b f(x)\, dx,$$

where $F(x)$ is a function such that $F'(x) = f(x)$. Thus we say that the value of the definite integral from a to b yields the area under the curve between a and b and the X-axis.

Illustration. Find the area bounded by $y = 2x$, the X-axis, and the ordinates at $x = 0$ and $x = 3$.

Solution.

$$\text{Area} = \int_0^3 2x\, dx$$
$$= \left[\frac{2x^2}{2}\right]_0^3$$
$$= 9.$$

The configuration is shown in Fig. 102. Check by the formula for the area of a triangle.

Illustration. Find the area of the ellipse:

$$\frac{x^2}{a^2} + \frac{y^2}{b^2} = 1.$$

Solution. The ellipse is symmetrical about the axes. We can therefore find the area in the first quadrant and multiply it by 4.

$$A_1 = \int_0^a y \, dx = \int_0^a \frac{b}{a} \sqrt{a^2 - x^2} \, dx$$

$$= \frac{b}{a} \left[\frac{x}{2} \sqrt{a^2 - x^2} + \frac{a^2}{2} \arcsin \frac{x}{a} \right]_0^a$$

$$= \frac{b}{a} \left[0 + \frac{a^2}{2} \arcsin 1 - 0 - \frac{a^2}{2} \arcsin 0 \right]$$

$$= \frac{b}{a} \left[\frac{a^2}{2} \cdot \frac{\pi}{2} \right] = \frac{ab\pi}{4}.$$

$$\therefore \quad \text{Area} = 4A_1 = 4 \left(\frac{ab\pi}{4} \right) = ab\pi.$$

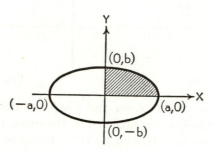

Fig. 103

147. Exercise XIV.

1. Find $\dfrac{dy}{dx}$.

    (a) $y = e^x \sin 3x^2 + 7\sqrt{x} - \arcsin \sqrt{x}$.

    (b) $y = 3x^2 - 6x + 3 \log_e 5x^2$.

    (c) $b^2x^2 + a^2y^2 = a^2b^2$.

    (d) $x^2 + y^2 + Dx + Ey + F = 0$.

    (e) $y = 2(x^{\frac{1}{2}} + x^{\frac{3}{2}}) \arctan \sqrt{x}$.

2. Find $\dfrac{d^2y}{dx^2}$.

    (a) $y = e^x [\sin x^2 - \cos x^2]$.

    (b) $y = 6x^3 - x^2 + x + 108$.

    (c) $b^2x^2 + a^2y^2 = a^2b^2$.

3. Find the maximum and minimum values of:

$$y = 4x^3 - 15x^2 - 18x + 9.$$

4. Trace the curve:

$$y = \frac{6x}{1 + x^2}.$$

5. If $s = v_1 t - a_1 t^2$, find the velocity and acceleration at the end of 4 seconds. Find the maximum distance if $v_1 = 600$ and $a_1 = 150$.

6. Integrate:
   (a) $\int (4x^3 - 6x^2 + 3) \, dx$.
   (b) $\int 3^x \, dx$.
   (c) $\int e^{x^2} x \, dx$.

7. Evaluate:
   (a) $\displaystyle\int_0^1 (9x^2 - 6x + 7) \, dx$.
   (b) $\displaystyle\int_{-1}^3 (4x^3 - 3x^2) \, dx$.
   (c) $\displaystyle\int_0^\pi \sin x \, dx$.

8. Find the area bounded by $y^2 = 4x$ and the line $x = 4$.

15

ADDITIONAL TOPICS

148. Introduction. Mathematics has often been referred to as the Queen of the Sciences. It is the trunk of the tree from which the other sciences emerge as branches. The value of mathematics in the fields of business, engineering, industry, and science has been well established. Daily observations emphasize the need to know as much about this subject matter as is individually practical. In the preceding chapters we have presented the basic fundamentals of a first-year college course for students who may not pursue the subject further. The retention of the basic principles and procedures will prove to be of some practical value. However, there is much more to mathematics than these basic principles.

There is the use of mathematics in *logic* and *deductive reasoning;* conversely, logic may be thought of as the core of any system of mathematics. Thus the study of mathematics should develop a deductive reasoning process. It should also be pointed out that more mathematics is being invented each day: that we create systems of mathematics and that each system is the building of a logical structure.

Mathematics is also concerned with *inductive reasoning,* which is a part of the so-called *scientific method.* In this method the scientist gathers empirical data, either from nature or from an experiment, and on the basis of these data infers a law or generalization. Conclusions are then predicted from the generalizations. If later observations do not confirm these predictions, the generalization must be discarded and a new one created.

In this chapter we shall present a few supplementary basic principles, some advanced topics, and a few problems which may mix practical applications with just "things."

149. Foundation. We shall not define mathematics but rather emphasize that the creation of a system of mathematics is the formation of a *logical structure,* which basically consists of primary state-

232

ments and secondary statements. *Primary statements* include undefined terms, defined terms, and assumptions that are accepted without proof. *Secondary statements* are the conclusions that are reached from the primary statements.

To start, then, we must have some undefined terms. In high-school geometry, a *point* was undefined. General mathematics usually starts with the term **set** or **class.** It might be said that these terms designate a group or a collection of objects with some common property, but then we should define group or collection. Instead, let us simply give some examples:

(a) The players on a baseball team form a set;

(b) all men over 35 years of age form a set;

(c) all the even numbers form a set.

In example (a) above, the players are called **members or elements** of the set. The word *distinct* is frequently used to indicate that an element occurs only once in a set. If a set has no elements, it is called a **null set.**

After the undefined terms have been chosen, other words are defined in terms of these and already defined quantities.*

Illustrations.

1. Let x represent an unspecified element of a set; then x is said to **vary** over the set and is called a **variable** on the set. Let the set be the positive integers; then x may be 1, 2, 3, etc.

2. If x varies over a set which consists of only one element, then x is called a **constant.** Let the set be the number 3; then $x = 3$.

3. An **ordered pair of elements** is a set which consists of two elements, one of which has been designated the first. We indicate this by **(a,b),** which also specifies that a has been ordered to be the first element and b the second.

In our definitions we include words of operation such as the process of *addition* or the process of *multiplication.* We also define relationships such as *congruent, equivalent, subsets, disjointed sets,* etc.

After having obtained a sufficient number of undefined and defined terms, we establish some basic assumptions which are statements about the terms; these are called *axioms* and *postulates.* They take the form of a complete sentence that makes an assertion which may be either true or false. However, they must be *consistent,* i.e., noncontradictory.

---

* In defining the mathematical terms we make use of our current language, and all such words are defined by a recognized dictionary.

Illustrations.

1. Two points determine a straight line.
2. All 30° angles are equal.
3. All college students are intelligent.

We accept the assumptions and then proceed to draw conclusions. These take the form of secondary statements. However, the *validity* of the conclusions must be established. We may illustrate the testing of a validity by means of a simple form of a deductive logical structure called a syllogism. A **syllogism** usually consists of three statements, the first two of which are the *premises* and the last of which is the *conclusion*.

Some very confusing statements may result in the use of syllogisms and point out the need for testing the validity of conclusions in order not to misuse the English language.

Illustration 1.

All college students are intelligent. ⎫
Mary is a college student.   ⎬ Hypothesis
Therefore, Mary is intelligent.   Conclusion

Solution. This is a valid conclusion since Mary belongs to a set which lies entirely within the group of intelligent people as per the hypothesis. Now, you may know Mary and not agree with the conclusion, but then you must find fault with the hypothesis and not the conclusion.

Illustration 2.

Some baseball players are left-handed. ⎫
Bill Jones is a baseball player.   ⎬ Hypothesis
Therefore, Bill Jones is left-handed.   ⎭ Conclusion

Solution. This conclusion is not valid because Bill Jones is a member of a set which is *bigger* than the set of left-handers, and thus he is not necessarily a member of the smaller set.

Mathematicians also create schemes and diagrams to aid man in his understanding of terminology and processes. One such scheme is to represent the elements of a set by a circle. Thus all the elements of a set K are inside the circle K, and all elements that are not in the set K are outside the circle. In Fig. 104 we have presented pictorially the following concepts:

a. *Set, K.* Fig. 104(a).
b. *Subset.* L is a subset of K. Fig. 104(b).
c. *Overlapping sets.* Fig. 104(c).

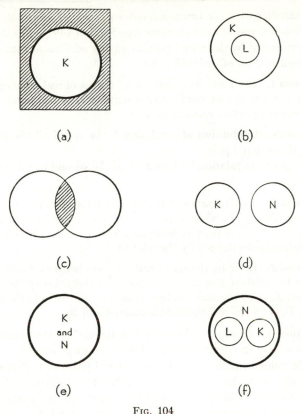

Fig. 104

d. *Disjointed sets.* K and N are disjointed sets. Fig. 104(d).

e. *Equivalent sets* may be represented by the same circle. Fig. 104(e).

f. *Disjointed subsets.* L and K are disjointed subsets of N. Fig. 104(f).

It is left for the student to construct the appropriate diagrams for testing the validity of the two syllogisms above.

150. Function. One of the most important terms in mathematics is the term *function*. In Section 16 we gave a very useful definition of the word function in terms of dependent and independent variables. It is desirable, however, to have a deeper appreciation of this word. Consequently, in this section we shall develop our thoughts about the word "function" along what has been called the modern trend. We are now familiar with the terms *set, element, variable, constant, ordered pair of elements,* and *distinct elements* (see Section 149). The development is continued by defining and illustrating additional concepts.

A **relation** is a set of ordered pairs.

Illustration 1. Let x and y be elements of the set of real numbers. The relation $x > y$ is the set of all ordered pairs (x,y) of real numbers such that $x - y$ is a positive real number.

The **domain of definition of a relation** is the set of all the first elements of the ordered pair.

The **range of the relation** is the set of all the second elements of the ordered pair.

Illustration 2. In Illustration 1 both the domain of definition and the range were the set of real numbers. If we let $x = 5$ and $y = 2$, then we see that $5 - 2 = 3$, a positive real number, so that $(5,2)$ belongs to the set of ordered pairs defined by the relation $x > y$.

Illustration 3. Let x be the set of real numbers between 2 and 10, and consider the relation $y = 2x$. The domain of definition of the relation is the set, X, of all the real numbers between 2 and 10, and the range is the set, Y, of all the real numbers between 4 and 20.

A **function** is a relation. However, this definition is not sufficiently precise to satisfy the deep thinker. Let us be more specific.

A **single-valued function** is a set of ordered pairs such that no two ordered pairs have the same first element. The *domain of definition of the function* is the set of all the first elements of the ordered pairs, and the *range of the function* is the set of all the second elements of the ordered pairs. If $\{x,y\}$ is the set of the ordered pairs, then in this definition for each x there is one and only one y. There is a trend to limit the definition of a function in this manner. However, we also have application for the concept of a multiple-valued function.

A **multiple-valued function** is a set of ordered pairs in which two distinct ordered pairs may have the same first element. Thus, if $\{x,y\}$ is the set of ordered pairs, then in this definition for each x there may exist one or more values of y.

We shall now state another form of the definition of a function:

A **functional relationship** exists between two variable sets if there is a correspondence between the elements of the two sets, these elements being arranged in some specified order.

Illustration 4. In Illustration 3 we gave an example of a single-valued function. The equation $y^2 = 4x$ is an example of a multiple-valued function if we let x be the variable element of the domain of definition of the function and y be the element of the range of the function.

There are many ways in which we may show a functional relationship:

I. A formula or an equation.
II. A graph.
III. A table of data.
IV. A statement of a principle.
V. A series.

Let f be a given function; X (with elements x), the domain of definition; and Y (with elements y), the range. For a given ordered pair (x,y) of f, we shall call x the **source** of y under f, and y the **image** of x under f or the **value of the function f at x**; and denote it by $f(x)$. Now if it is possible to obtain a function from f by interchanging in each ordered pair (x,y) the places of x and y, we call this function the **inverse function of f**. It should be clear that its domain of definition is Y and its range is X. Some authors make a distinction between the inverse function which is obtained from the set of ordered pairs, no two of which have the same second element, and the set of ordered pairs obtained by interchanging x and y in f when f is such that two or more of its ordered pairs have the same second element. The latter case is then called an **inverse relation**.

If the single-valued function f has a single-valued inverse f^{-1} relating X and Y, then f and f^{-1} are said to constitute a **one-to-one correspondence** between X and Y.

Illustration 5. Let f be given by $y = 2x$, as in Illustration 3; then for each value of x there is one and only one value of y. Now f^{-1} is $x = \frac{1}{2}y$, and for each value of y there is one and only one value of x.

151. The Binary Number System. Our number system is an example of a mathematical system based on a deductive logical structure. The present-day system has a basis of 10, and we speak of *units, tens, hundreds, thousands*, etc. The common logarithms use 10 as a base. This number system is readily adapted to our decimal system and powers of tens expressions. We can write a number in various ways by simply moving the decimal point and using powers of 10; **for** example:

$$1956 = 195.6 \times 10 = 19.56 \times 10^2 = 1.956 \times 10^3.$$

We also know that a number is a quantity which is represented by a group of *digits*. Thus 1956 is a number made up of the *integers* 1, 9,

5, and 6. Mathematicians are a little more precise, and we express a number symbolically by:

$$N = d_0 + d_1R + d_2R^2 + d_3R^3 + \cdots$$

where d_i are digits and R is called the **radix**. In our usual number system R is 10. Thus:

$$1956 = 6 + 5(10) + 9(10)^2 + 1(10)^3$$
$$= 6 + 50 + 900 + 1000.$$

However, we write the number in reverse order so that **in general**:

$$N = \cdots d_3d_2d_1d_0$$

represents the number.

Although the above is the accepted method for writing numbers, there are others. The student is undoubtedly familiar with the Roman number system where:

$$
\begin{array}{ll}
I = 1 & L = 50 \\
V = 5 & C = 100 \\
X = 10 & D = 500 \\
& M = 1000
\end{array}
$$

The number 128 is written:

$$
\begin{array}{rl}
100 = & C \\
10 = & X \\
10 = & X \\
5 = & V \\
\underline{3 =} & III \\
128 = & CXXVIII
\end{array}
$$

The system we shall now consider is the one which chooses a radix of two and is called the **binary** system. Here we use only two symbols, 0 and 1. The numbers are shown in the table on p. 239.

As we said, the radix is 2, and we write:

$$2^0 = 1, \quad 2^1 = 10, \quad 2^2 = 100, \quad 2^3 = 1000, \quad 2^4 = 10000, \text{ etc.}$$

Thus the number:

$$137 = 128 + 8 + 1 = 2^7 + 2^3 + 2^0 = 10001001.$$

The arithmetic of this number system is not easy to do with pencil and paper. We have the following addition tables:

$$0 + 0 = 0,$$
$$0 + 1 = 1.$$

$1 + 1 = 0$ with 1 carry.

Decimal	Binary	Decimal	Binary
0	0	11	1011
1	1	12	1100
2	10	13	1101
3	11	14	1110
4	100	15	1111
5	101	16	10000
6	110	32	100000
7	111	64	1000000
8	1000	128	10000000
9	1001	100	1100100
10	1010	1000	1111101000

Illustration. Add 3 and 7 in the binary system.
Solution.

$$
\begin{array}{rl}
3 = & 0011 \\
7 = & 0111 \\
\hline
\text{First add.} \quad = & 0100 \\
\text{carry} \quad = & 11 \\
\hline
\text{Second add.} \quad = & 0010 \\
\text{carry} \quad = & 1 \\
\hline
\text{Sum} \quad = & 1010
\end{array}
$$

The multiplication table is very simple.

Multiplicand

		0	1
Multiplier	0	0	0
	1	0	1

To multiply, however, is by no means simple. Although:

2	10
×	11
3	10
‖	
6	10
6 =	110

looks easy, we see that $11 \times 13 = 143$:

$$
\begin{array}{r}
1011 \\
1101 \\
\hline
1011 \\
0000 \\
1011 \\
1011 \\
\hline
143 = 10001111 = 2^7 + 15
\end{array}
$$

may give us trouble in the addition.

The student may find it interesting to play with some arithmetic problems.

152. The Electronic Brain. The binary number system finds application in one of the most fascinating industries to emerge from World War II, namely, the production of electronic calculators. These have been referred to as "giant brains," "calculating monsters," etc. Names such as the "Eniac," "Illiac," "Univac," "Maddida," and "Maniac" have become general accepted terminology when speaking of specific machines. Today there are some 165 manufacturers of medium or large-scale calculating machines. The machines cost anywhere from a few thousand dollars to over a million dollars.

One of the original calculators was undoubtedly a version of the Japanese *abacus*, which is still in use today. It has been traced back to the Tigris-Euphrates Valley some 5000 years ago. The most prevalent calculator in the United States is the slide rule, which is based on the principle of logarithms.* There are a number of electric desk calculators available, and the engineering profession has found such a great need for more precision that they may soon replace the slide rule as standard equipment for engineers. Mathematics has indeed kept pace, and one of the most fascinating branches of mathematics is the one known as *numerical analysis*.**

The use of large-scale digital calculators is becoming so prevalent that big business is thinking of automatic production and offices. Most calculators are based on the fundamental mathematical principle called the *binary number system* (see Section 151).

---

* See C. C. Bishop, *Slide Rule* (New York: Barnes & Noble, Inc., 1956).
** See K. L. Nielsen, *Methods in Numerical Analysis* (New York: The Macmillan Co., 1956).

153. Evaluation of Higher-Order Determinants.   The theory of determinants is an interesting example of developing a group of definitions, notations, schemes, and logical processes with applications. We have already discussed second- and third-order determinants in Sections 62 and 63. We shall now develop the evaluation of higher-order determinants, and to do so we rely on a number of important properties of determinants. It should be emphasized that these properties hold for determinants of any order. We shall assume the following two properties:

I. *The value of the determinant is not changed if the elements of any row (or column) are multiplied by any quantity and added to the corresponding elements of another row (or column).*

II. *The value of a determinant is the sum of the results obtained by multiplying the elements of any row (or column) by their cofactors.*

To understand property II we need to define a minor and a cofactor.

The **minor** of any element of a determinant is defined as the determinant of next lower order, obtained by striking out the row and column in which the element stands.

Thus the minor of a_2 in the following determinant:

$$\begin{vmatrix} a_1 & b_1 & c_1 \\ a_2 & b_2 & c_2 \\ a_3 & b_3 & c_3 \end{vmatrix} \quad \text{is} \quad \begin{vmatrix} b_1 & c_1 \\ b_3 & c_3 \end{vmatrix}.$$

The minor of b_2 is $B_2 = \begin{vmatrix} a_1 & c_1 \\ a_3 & c_3 \end{vmatrix}.$

The **cofactor** of an element is equal to the minor of the element if the sum of the number of the row and the number of the column in which the element stands is an even number; otherwise it is the negative of its minor.

We can illustrate property II.

Illustration.   Expand:

$$D = \begin{vmatrix} a_1 & b_1 & c_1 \\ a_2 & b_2 & c_2 \\ a_3 & b_3 & c_3 \end{vmatrix}$$

by use of minors.

Solution.

$$D = a_1 \begin{vmatrix} b_2 & c_2 \\ b_3 & c_3 \end{vmatrix} - a_2 \begin{vmatrix} b_1 & c_1 \\ b_3 & c_3 \end{vmatrix} + a_3 \begin{vmatrix} b_1 & c_1 \\ b_2 & c_2 \end{vmatrix}$$

or

$$D = -b_1 \begin{vmatrix} a_2 & c_2 \\ a_3 & c_3 \end{vmatrix} + b_2 \begin{vmatrix} a_1 & c_1 \\ a_3 & c_3 \end{vmatrix} - b_3 \begin{vmatrix} a_1 & c_1 \\ a_2 & c_2 \end{vmatrix}$$

To evaluate a higher-order determinant, reduce all the elements of a row (or column) except one to zero and expand the determinants by minors of the elements of that row (or column).

Illustration. Evaluate:

$$D = \begin{vmatrix} 3 & -2 & 1 & -1 \\ -1 & 1 & 2 & -2 \\ 2 & 0 & 3 & 1 \\ 4 & 3 & 4 & -3 \end{vmatrix}$$

Solution. Let us choose to make all the elements of column "2" zero except the element in the second row.

a. Multiply the elements of the second row by 2 and add to the elements of the first row.
b. Leave second and third row unchanged.
c. Multiply elements of second row by -3 and add to elements of fourth row.

We obtain:

$$D = \begin{vmatrix} 1 & 0 & 5 & -5 \\ -1 & 1 & 2 & -2 \\ 2 & 0 & 3 & 1 \\ 7 & 0 & -2 & 3 \end{vmatrix}$$

Expand in terms of the elements of the second column.

$$D = -0 \begin{vmatrix} \ \end{vmatrix} + 1 \begin{vmatrix} 1 & 5 & -5 \\ 2 & 3 & 1 \\ 7 & -2 & 3 \end{vmatrix} - 0 \begin{vmatrix} \ \end{vmatrix} + 0 \begin{vmatrix} \ \end{vmatrix}$$

$$= \begin{vmatrix} 1 & 5 & -5 \\ 2 & 3 & 1 \\ 7 & -2 & 3 \end{vmatrix}$$

d. Add the third column to the second.
e. Multiply the elements of the first column by 5 and add to the elements of the third column.

$$D = \begin{vmatrix} 1 & 0 & 0 \\ 2 & 4 & 11 \\ 7 & 1 & 38 \end{vmatrix}$$

Expand in terms of the elements of the first row.

$$D = 1 \begin{vmatrix} 4 & 11 \\ 1 & 38 \end{vmatrix} - 0 \begin{vmatrix} \ \end{vmatrix} + 0 \begin{vmatrix} \ \end{vmatrix} = 4(38) - 1(11) = 141.$$

154. Newton's Method for Finding Roots. An interesting example of a formal procedure in a mathematical system is the procedure of *successive approximations*. We shall illustrate it by a new method for finding the irrational roots of an algebraic equation. (See Section 104 for Horner's method.) The method is based on Newton's formula for approximating the root:

$$x_{i+1} = x_i - \frac{f(x_i)}{f'(x_i)}.$$

where $f'(x_i)$ is the value of the derivative of $f(x)$ at $x = x_i$. We shall illustrate the procedure with an example.

Illustration. Find the real root of:

$$x^3 - x^2 + 4x - 5 = 0.$$

Solution.

$$f(x) = x^3 - x^2 + 4x - 5 = 0.$$
$$f'(x) = 3x^2 - 2x + 4.$$

From a graph of the function we see that there is a root between 1 and 2. We thus choose the first value of x to be 1 and calculate:

$$f(1) = -1; \quad f'(1) = 5.$$

Then:

$$x_2 = 1 - \frac{-1}{5} = 1 + \frac{1}{5} = 1.2.$$

Now calculate:

$$f(1.2) = .088 \quad \text{and} \quad f'(1.2) = 5.92.$$

Then:

$$x_3 = 1.2 - \frac{.088}{5.92} = 1.2 - .015 = 1.185.$$

The values of the function and the derivative may be found by synthetic division. It is advantageous to arrange the work in a table.

i	x	$f(x)$	$f'(x)$
1	1	-1	5
2	1.2	.088	5.92
3	1.185		

155. Dimensional Analysis. In applying mathematical formulas to the solution of problems which arise from the physical sciences, it is essential that uniform dimensions be used. These formulas should always be checked for their dimensional accuracy by using ratios and considering the expressions for the dimensions as algebraic quantities.

Illustration 1. Change 7 yards to inches.

Solution. We have the following defined relations:

$$1 \text{ yard} = 3 \text{ feet} \quad \text{and} \quad 1 \text{ foot} = 12 \text{ inches.}$$

If we write these in ratio form we see that:

$$\frac{3 \text{ ft.}}{1 \text{ yd.}} = 1 \quad \text{and} \quad \frac{12 \text{ in.}}{1 \text{ ft.}} = 1.$$

Thus we may multiply any quantity by such expressions without changing its value. Consequently:

$$7 \text{ yd.} = 7 \text{ yd.} \left(\frac{3 \text{ ft.}}{1 \text{ yd.}}\right) = 21 \text{ ft.} = 21 \text{ ft.} \left(\frac{12 \text{ in.}}{1 \text{ ft.}}\right) = 252 \text{ in.}$$

Some dimensions are given in terms of the word "per," e.g., feet per second. In this sense the word "per" means division or a ratio and is frequently indicated by a fractional bar, e.g., ft./sec.

Illustration 2. How many feet per second does an automobile travel if its speed is 50 miles per hour?

Solution.

$$x \frac{\text{ft.}}{\text{sec.}} = \frac{50 \text{ mi.}}{\text{hr.}} = \frac{50 \text{ mi.}}{1 \text{ hr.}} \times \frac{1 \text{ hr.}}{3600 \text{ sec.}} \times \frac{5280 \text{ ft.}}{1 \text{ mi.}}$$

$$= \frac{50 \text{ mi.}}{\text{hr.}} \times \frac{\text{hr.}}{3600 \text{ sec.}} \times \frac{\overset{220}{\cancel{5280}} \text{ ft.}}{\text{mi.}}$$

$$= \frac{220 \text{ ft.}}{3 \text{ sec.}} = 73\tfrac{1}{3} \text{ ft./sec.}$$

Illustration 3. The drag on a projectile moving in the air may be expressed by the formula:

$$D = K_D \rho_a d^2 V^2,$$

where D is the drag;   (mass)(ft.)/(sec.)²

     K_D is the drag coefficient;   dimensionless

     ρ_a is the air density;   mass/(ft.)³

     d is the diameter of projectile; ft.

     V is the projectile velocity relative to the air; ft./sec.

Prove that K_D is dimensionless.

Solution.

$$D = K_D \rho_a d^2 V^2.$$

$$\frac{(\text{m.})(\text{ft.})}{(\text{sec.})^2} = K_D \frac{\text{m.}}{(\text{ft.})^3} \times \frac{(\text{ft.})^2}{1} \times \frac{(\text{ft.})^2}{(\text{sec.})^2}$$

$$= K_D \frac{(\text{m.})(\text{ft.})}{(\text{sec.})^2}.$$

Both sides of the equation now have the same dimensions providing K_D is dimensionless.

156. Problems for Fun. We shall close the text of this book with four problems for fun. Each problem illustrates a procedure or concept in problem solving, and it is hoped that they may prove interesting.

I. You Forgot Something. Here is an interesting development from elementary algebra.

Let:	$a = b.$
Multiply by a:	$a^2 = ab.$
Subtract b^2:	$a^2 - b^2 = ab - b^2.$
Factor:	$(a - b)(a + b) = b(a - b).$
Divide by $a - b$:	$a + b = b.$
Substitute $a = b$:	$b + b = b,$
or:	$2b = b.$
Divide by b:	$2 = 1.$ What happened?

II. How Old Are You? Try this on your friends.

1. Take your age and double it.
2. Add 5 to the results.
3. Multiply the result by 50.
4. Add the amount of change you have less than one dollar.
5. Subtract the number of days in a year (365).
6. Tell me the resulting number.
7. Your age is . . .; the amount of change you have is . . ..

By adding 115 to the number obtained in (6) the age will appear as the first two digits and the amount of change will appear as the last two digits in the results.

Proof. Let the age be x and the amount of change be y. Then the operations listed above yield:

$$50(2x + 5) + y - 365 = N$$

or

$$100x + y = N + 115 = d_1 d_2 d_3 d_4.$$

Since $y < 100$ and x is multiplied by 100, we have:

$$x = d_1 d_2 \quad \text{and} \quad y = d_3 d_4.$$

Try it!

III. On What Day of the Week Were You Born? Consider the formula:

$$N = d + 2m + \left[\frac{3(m+1)}{5}\right] + Y + \left[\frac{Y}{4}\right] - \left[\frac{Y}{100}\right] + \left[\frac{Y}{400}\right] + 2,$$

where:

d is the day of the month;

m is the number of the month in the year, with the qualification that January and February are counted as the 13th and 14th months of the preceding year;

Y is the year;

and the bracket expression means to take the largest **integer** *not greater than the enclosed number.*

If the N be divided by 7, the remainder will be the day of the week of a given date where Sunday is counted as the first day. A remainder of 0 indicates Saturday.

Illustration.   On what day of the week did January 1, 1956, fall?

Solution.   The date January 1, 1956, must be written 13/1/1955 according to the definition of m given above. Then we have $d = 1$, $m = 13$, $Y = 1955$, and:

$$N = 1 + 26 + \left[\tfrac{42}{5}\right] + 1955 + \left[\tfrac{1955}{100}\right] - \left[\tfrac{1955}{100}\right] + \left[\tfrac{1955}{400}\right] + 2$$
$$= 1 + 26 + 8 + 1955 + 488 - 19 + 4 + 2$$
$$= 2465$$
$$= 352(7) + 1.$$

Therefore, January 1, 1956, fell on Sunday.

Verify the formula for the present day and then calculate the day of the week for your birthday.

IV. The Ladder Problem.

Given Fig. 105. Find x.

Solution.

By the Pythagorean Theorem we have:

(1) $30^2 = x^2 + w^2$.

(2) $20^2 = x^2 + v^2$.

By similar triangles we have:

$$\frac{w}{10} = \frac{x}{r} \quad \text{or} \quad w = \frac{10x}{r}.$$

$$\frac{v}{10} = \frac{x}{x-r} \quad \text{or} \quad v = \frac{10x}{x-r}.$$

By substitution we then obtain:

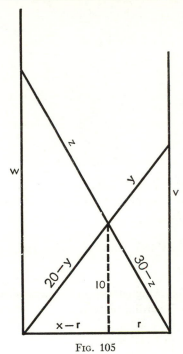

FIG. 105

$$(3) \begin{cases} 30^2 = x^2 + \left(\dfrac{10x}{r}\right)^2 \\ 20^2 = x^2 + \left(\dfrac{10x}{x-r}\right)^2. \end{cases}$$

For algebraic convenience we scale the problem by a factor of 10; i.e., divide each quantity by 10:

$$\frac{30}{10} = 3, \quad \frac{20}{10} = 2, \quad \frac{10}{10} = 1, \quad \frac{x}{10} = x_1, \quad \frac{r}{10} = r_1.$$

$$(4) \begin{cases} 9 = x_1^2\left(1 + \dfrac{1}{r_1^2}\right). \\ 4 = x_1^2\left(1 + \left[\dfrac{1}{x_1 - r_1}\right]^2\right). \end{cases}$$

The solution of this system of equations may be accomplished by solving the first equation for r_1 and the second for $(x_1 - r_1)^2$.

$$r_1 = \frac{x_1}{\sqrt{9 - x_1^2}} \quad \text{and} \quad (x_1 - r_1)^2 = \frac{x_1^2}{4 - x_1^2}.$$

From the first equation, then, we have:

$$(x_1 - r_1)^2 = \left[x_1 - \frac{x_1}{\sqrt{9 - x_1^2}} \right]^2.$$

Thus:

(5) $$\frac{x_1^2}{4 - x_1^2} = \frac{x_1^2}{9 - x_1^2} \left[\sqrt{9 - x_1^2} - 1 \right]^2$$

which yields one equation in one unknown. Let $n = x_1^2$ and simplify (5):

$$2(4 - n)\sqrt{9 - n} = 31 - 13n + n^2.$$

Square both sides and simplify:

$$n^4 - 22n^3 + 163n^2 - 454n + 385 = 0.$$

The smallest positive root is $n = 1.515818$.

$$\therefore \quad x_1 = \sqrt{n} = 1.231186 \quad \text{and} \quad x = 12.31186.$$

157. Exercise XV.

1. Draw all the following curves on the same sheet of graph paper, using only one set of coordinate axes:

    (a) $(x + 3)^2 + (y - 4)^2 = 1$.

    (b) $(x - 3)^2 + (y - 4)^2 = 1$.

    (c) $(x + 3)^2 + (y - 4)^2 = 0$.

    (d) $(x - 3)^2 + (y - 4)^2 = 0$.

    (e) $\dfrac{4x^2}{9} + \dfrac{y^2}{16} = 1$ for values $y < 2$.

    (f) $x^2 = -9(y + 5)$ for $-3 \leq x \leq 3$.

    (g) $\dfrac{(x - 11)^2}{1} + \dfrac{y^2}{9} = 1$.

    (h) $\dfrac{(x + 11)^2}{1} + \dfrac{y^2}{9} = 1$.

    (i) $y = 6$ for $-4 \leq x \leq -2$ and $2 \leq x \leq 4$.

    (j) $x = \sin \pi y$ for $10 \leq y \leq 12$.

    (k) $r = 5 \cos^2 \theta$ with the pole at $(0, -10)$ and the polar axis along $y = -10$.

    (l) $x^2 - 4(y + 6)^2 = 9$ for $-4 \leq x \leq 4$.

    (m) $x^2 + y^2 = 100$.

EXAMINATIONS

One of the best ways to review for an examination is to take a pre-examination. Consequently, to test your knowledge of the subject matter in this book we have prepared a set of examinations. There are three groups of examinations:

Part I — One-Hour Examinations.
Part II — Miscellaneous Examinations.
Part III — Final Examinations.

The examinations in Part I cover specific subject matter, and we have devoted one examination to each chapter. The examinations in Part II cover particular subjects such as algebra, graphing, trigonometry, etc. and are of varying lengths. The final examinations cover the entire book.

Most of the problems which have been assembled to form these examinations come from the author's file on examination questions and thus are from actual examinations that were given by the author during his college teaching career.*

Part I — One-Hour Examinations

Test #1. (Covers material in Chapter 1.)

1. Simplify:

$$\frac{1}{1 - \dfrac{1}{1 + x}} - \frac{1}{\dfrac{1}{1 - x} - 1}.$$

2. Combine into a single fraction:

$$\frac{a}{(a - b)(a - c)} + \frac{b}{(b - a)(b - c)} + \frac{c}{(c - a)(c - b)}.$$

---

* Some of the problems may have their origin in some book but have long since lost their identity to the author. If he has inadvertently "borrowed" a problem from another author, he hereby asks forgiveness in the interest of education.

3. Simplify:

$$\left[x^{\frac{1}{2}} \left\{ \left(\frac{x^{\frac{1}{2}}}{x^{\frac{1}{3}}} \right)^7 \right\}^{\frac{3}{2}} \right]^{\frac{4}{3}}.$$

4. Rationalize the denominator and simplify:

$$\frac{\sqrt{\frac{2}{3}} + \sqrt{\frac{3}{2}}}{\sqrt{\frac{2}{3}} - \sqrt{\frac{3}{2}}}.$$

5. Multiply $-\frac{1}{2} + \frac{1}{2}i\sqrt{3}$ by $-\frac{1}{2} - \frac{1}{2}i\sqrt{3}$.

Test #2. (Covers material in Chapter 2.)

1. The formula $C = \frac{5}{9}(F - 32)$ converts F degrees Fahrenheit into C degrees Centigrade.

   (a) Express F in terms of C.
   (b) Find F when $C = 100°$.
   (c) Find C when $F = 95°$.

2. Plot the graph of the function:

$$y = 6x - 3$$

   and find its zeros.

3. Find the distance between the points:

$$P_1(3,-2) \text{ and } P_2(6,2).$$

4. Find the values of each function by use of a table:

   (a) $\sin (-162° \, 17')$;
   (b) $\tan 212° \, 13'$;
   (c) $\cos (-375° \, 10')$.

5. Evaluate $\sin (\arctan \frac{5}{12})$.

Test #3. (Covers material in Chapter 3.)

1. Solve the equation:

$$\frac{3}{x-1} - \frac{2}{2-2x} = \frac{4}{\sqrt{3}}.$$

2. Prove the identity:

$$\sin^2 x + \cos^2 x = 1.$$

3. Prove the identity:

$$\tan \frac{x}{2} = \cot \frac{x}{2} - 2 \cot x.$$

4. Solve for x for values less than $360°$:

$$\cos \tfrac{1}{2}x = \tfrac{1}{2}.$$

5. If $\frac{1}{5}$ of a man's monthly income is deducted for taxes, $\frac{1}{10}$ is deducted for security benefits, $\frac{1}{3}$ is spent on rent and board, $\frac{1}{11}$ of the remainder is spent for clothes and transportation, and \$208 is left over, what is his actual income?

Test #4. (Covers material in Chapter 4.)

1. Find the first five terms in the expansion of:
$$(x^{\frac{1}{2}} - 3y^{-1})^{-3}.$$

2. Find the equation relating H, x, and y if H varies directly as x and inversely as the square of y, and if $H = 100$ when $x = 5$ and $y = \frac{1}{2}$.

3. Find the sixth term in the expansion of:
$$(x^3 - 2y^{-2})^{-1}.$$

4. Find $\sqrt[3]{\sqrt{2} + \sqrt{2}i}$.

5. Find the value of $(-\frac{1}{2} + \frac{1}{2}i\sqrt{3})^3$.

Test #5. (Covers material in Chapter 5.)

1. Given the points $P_1(-3,1)$, $P_2(5,-4)$, and $P_3(2,9)$. Find:

    (a) the length of the line segment P_1P_2;

    (b) the coordinates of the midpoint of the line segment P_1P_2;

    (c) the slope of the line through P_1 and P_2;

    (d) an equation of the line through P_1 and P_2;

    (e) an equation of a line through P_1 and perpendicular to the line P_1P_2;

    (f) the distance from P_3 to the line P_1P_2;

    (g) the rate of change of the linear function defining the line between P_2 and P_3;

    (h) the tangent of the angle from the line P_1P_2 to the line P_2P_3;

    (i) two other points each on the lines between P_1P_2, P_2P_3, and P_1P_3.

    (j) Draw the entire configuration.

2. Solve the system:
$$\begin{cases} u + v + 4x = 1, \\ v - u - 2 = -2x, \\ 2u - v - 3x = -3. \end{cases}$$

Test #6. (Covers material in Chapter 6.)

1. Solve the following quadratic equation:

$$9x^2 - 6x + 2 = 0.$$

2. Solve the following system of quadratic equations:

$$\begin{cases} x^2 + 2y^2 + 5 = 0, \\ 3x^2 + 18y + 3 = 0. \end{cases}$$

3. Find the equation of the circle through the points $P_1(-2,0)$, $P_2(0,4)$, and $P_3(1,1)$.

4. Find the equation of the line through the origin and perpendicular to the radical axis of the circles $x^2 + y^2 + 2x - 4y = 0$ and $x^2 + y^2 - 4x + 2y = 0$. Draw the configuration.

5. Determine the quadratic equation which has the sum of its roots equal to 7 and the product of its roots equal to 14.

Test #7. (Covers material in Chapter 7.)

1. Draw the parabola $2x^2 - 9y = 0$; find the length of the latus rectum, and locate the vertex, focus, axis, and directrix.

2. Given the hyperbola $64x^2 - 36y^2 = 2304$. Find the equation of its conjugate hyperbola; the vertices, foci, equations of the asymptotes, and directrices of both; and draw the complete configuration.

3. Simplify and put into standard form:

$$3x^2 + 2xy + 3y^2 + 6x - 14y + 19 = 0.$$

Test #8. (Covers material in Chapter 8.)

1. Calculate a table of values for the curve $r + 2 \cos \theta - 4 \sin \theta = 0$ and draw its graph.

2. Calculate a table of values for the curve $r = a \sin 2\theta$ and draw its graph.

3. Change the equation $r + 2 \sin \theta = 0$ to rectangular coordinates. Name the curve.

4. Change the equation $9x^2 + 16y^2 = 144$ to polar coordinates. Find the eccentricity and name the curve.

5. Plot the curve $2r \cos \theta + 3r \sin \theta = 6$.

Test #9. (Covers material in Chapter 9.)

1. Given the progression $\frac{1}{2}, \frac{5}{6}, \frac{7}{6}, \cdots$

   (a) What kind of progression is it?
   (b) Find the value of the seventh term.
   (c) Find the sum of the first seven terms.

2. Change the repeating decimal .2565656 . . . into an equivalent common fraction.
3. How many permutations can be made of the letters in the word *mathematics* when taken all at a time?
4. How many committees of 5 people can be selected from a group of 10 people?
5. One hundred tickets are sold at a dollar each. The winning ticket pays $75.00. What is the value of the expectation of a single draw?

Test #10. (Covers material in Chapter 10.)

1. Find all the roots of:
$$2x^4 + 5x^3 - 10x^2 + 5x - 12 = 0.$$

2. Prove the Remainder Theorem.
3. Solve algebraically the cubic equation:
$$4x^3 - 12x^2 - x + 3 = 0.$$

4. Given the equation:
$$x^6 - 3x^5 - 4x^4 + 3x^3 - 6x^2 - x - 1 = 0.$$

Test the equation for positive and negative roots.
List all the possible rational roots.
Find $f(\pm 1)$ and $f(\pm 2)$.
Plot the function in the range $-3 \leq x \leq 3$.
5. Find the real root correct to 2 decimal places by use of Horner's method:
$$28x^3 - 11x^2 + 15x - 2 = 0.$$

Test #11. (Covers material in Chapter 11.)

A table of logarithms is needed.

1. Calculate the value of N if:
$$N = \frac{3.142\sqrt{2.79}}{(0.9342)(1.320)}.$$

2. Find the value of ln 1.42.
3. Find the value of x for which:
$$\log (x - 5) + \log 5x = 1.$$

4. If log 2 = 0.3010 and log 3 = 0.4771, find:
$$\log \sqrt{\tfrac{27}{8}}.$$

5. Prove that:

$$\log_a \frac{M}{N} = \log_a M - \log_a N.$$

Test #12. (Covers material in Chapter 12.)

A table of logarithms is needed.

1. Given $a = 3$, $\angle ABC = 30°$, $\angle BCA = 90°$ in $\triangle ABC$; find the remaining parts of the triangle and the area.
2. Solve the triangle ABC if:

$$a = .429, \quad c = .634, \quad \text{and} \quad \beta = 78° \; 14'.$$

3. Find the area in $\triangle ABC$ if:

$$a = 12, \quad b = 18, \quad \text{and} \quad c = 25.$$

4. Prove the law of sines for an acute angle α.
5. A vector of magnitude 5 is elevated at 30° with the horizontal. Find its vertical and horizontal components.

Test #13. (Covers material in Chapter 13.)
1. Find the distance between the points:

$$P_1(3,-1,5) \quad \text{and} \quad P_2(-1,4,2).$$

2. Find the center and radius of the sphere:

$$x^2 + y^2 + z^2 - 4x + 6y + 12z = 0.$$

3. Find the direction numbers of the line between:

$$P_1(4,1,3) \quad \text{and} \quad P_2(2,7,1).$$

4. Given the plane:

$$3x - y + 2z = 6.$$

Find the intercepts on the axes, find the equations of the traces, and sketch the plane.
5. Find the symmetric form of the equations of the line between the points P_1 and P_2 of problem 3.

Test #14. (Covers material in Chapter 14.)
1. Find $\frac{dy}{dx}$ for:
   (a) $y = 4x^3 - 3x^2$.
   (b) $xy = y + x$.
2. Integrate:

(a) $\int \left(x^2 + \dfrac{1}{x^2} \right) dx.$

(b) $\int \sin x^2\, x\, dx.$

3. Find the area under the curve $y = x^3$ from $x = 0$ to $x = 3$.

4. Trace the curve:
$$y = 3x - x^3.$$

5. A particle falls according to the formula $s = 16t^2$. What is its speed at the end of 3 seconds? What is its acceleration at the end of 2 seconds?

Part II — Miscellaneous Examinations

A. An Examination on College Algebra. Time: 3 hours.

Answer any ten questions.

1. Simplify:

(a) $\left(\dfrac{x}{x - y} - \dfrac{x}{x + y} \right) \div \left(\dfrac{x + y}{x - y} + \dfrac{x - y}{x + y} \right).$

(b) $\dfrac{x + 3y^{-1}}{x^2 + 9y^{-2}}.$

2. Solve for z in the following system of equations:
$$\begin{cases} x + 2y + 3z + w = 3, \\ 2x + y + 2z - 3w = 15, \\ x - 2y - 2z + w = 1, \\ 3x - 3y + z + 2w + 4 = 0. \end{cases}$$

(Hint: Eliminate w, then use 3rd-order determinants.)

3. (a) Solve for x in $x^2 + 2x + 5 = 0$.
   (b) Find k so that the zeros of $x^2 - 2kx + 2 - k$ are equal.

4. Find the common solutions of $2x^2 - y^2 = 68$ and $x - y = 4$. Sketch the graphs.

5. Given $\log 2 = 0.3010$, $\log 3 = 0.4771$, and $\log 7 = 0.8451$. Find:

(a) $\log \dfrac{27\sqrt{14}}{28}$;   (b) $\log_3 2$.

6. The horsepower that a shaft can safely transmit varies directly as its speed and the cube of its diameter. If a 3 inch shaft making 100 revolutions per minute can transmit 85 horsepower, what horsepower can a 4 inch shaft transmit at 150 revolutions per minute?

7. (a) Find the sixth term of $\left(x - \dfrac{1}{2x}\right)^{11}$.

    (b) Expand and simplify $(1 - i)^5$.

8. (a) Find values of x for which $9x^2 - 2 > 15x + 4$.

    (b) Find values of x for which $9x^2 - 2 < 15x + 4$.

9. (a) What rational fraction is equivalent to $0.1272727\ldots$?

    (b) Insert 5 arithmetic means between 2 and 28.

10. Find all the roots of $x^4 + x^3 - 3x^2 - 3x = 0$. Sketch the graph.

11. Locate the real roots of $x^3 + 2x + 5 = 0$ and find one root accurately to two decimal places.

12. (a) In how many ways can you choose ten questions to answer on this examination (assuming equal knowledge of all)?

    (b) What is the probability of throwing a 7 in one toss of two dice? Of throwing an 11 in one toss? Of throwing a 7 or an 11 in one toss?

13. The sum of the squares of two numbers is 41 and their product plus the square of the smaller is 36. Find the numbers.

B. An Examination on Graphing. No time limit.

I. Plot the graphs of the following equations on rectangular coordinate paper.

  1. $y = 6 - x - x^2$.   5. $y = x^2 - 4x + 4$.

  2. $y = \pm\sqrt{16 - 4x^2}$.   6. $y = x^3 + x - 4$.

  3. $y = x^3 - x^2 + 7$.   7. $y = \pm\frac{1}{2}\sqrt{9 + x^2}$.

  4. $y = \pm\sqrt{4 + x^2}$.   8. $y = \pm\frac{2}{3}\sqrt{x^2 - 9}$.

  9. $x^{\frac{1}{2}} + y^{\frac{1}{2}} = 2$.

  10. $\frac{1}{2}(3x - 10) - 3(1 - \frac{1}{4}y) = \frac{1}{4}(9y - 20)$.

II. Plot the graphs of both loci on the same rectangular coordinate paper.

  1. $\begin{cases} y = \frac{1}{2}(4x - 7), \\ y = 5 - 3x. \end{cases}$ 5. $\begin{cases} xy = 6, \\ x^2 + y^2 = 25. \end{cases}$

  2. $\begin{cases} 2x + 3y = 7, \\ xy + y^2 = 5. \end{cases}$ 6. $\begin{cases} x + y = 3, \\ xy = -3. \end{cases}$

  3. $\begin{cases} x^2 + y^2 = 9, \\ 9x^2 + 16y^2 = 144. \end{cases}$ 7. $\begin{cases} 4x^2 + y^2 = 61, \\ 2x - 1 = y. \end{cases}$

  4. $\begin{cases} x^2 + 2xy = 5, \\ x^2 - xy + y^2 = 3. \end{cases}$ 8. $\begin{cases} x^2 - 10y = 19, \\ x^2 - 4y^2 = 5. \end{cases}$

9. $\begin{cases} 2x^2 - y^2 = 4, \\ 3x^2 + 4y^2 = 12. \end{cases}$

10. $\begin{cases} x^2 + 4y^2 = 16, \\ 16y^2 - x^2 = 16. \end{cases}$

11. $\begin{cases} xy = 7, \\ 4x^2 - 9y^2 + 36 = 0. \end{cases}$

12. $\begin{cases} 4x^2 + 9y^2 = 36, \\ 3x + 2y = 0. \end{cases}$

13. $\begin{cases} xy = -12, \\ x^2 - 2y^2 + 2 = 0. \end{cases}$

14. $\begin{cases} x - \frac{3}{2}(y + 1) = 1, \\ y - \frac{1}{5}(x + 3) = 1. \end{cases}$

15. $\begin{cases} y^2 = 12x, \\ 9x^2 - 16y^2 = 144. \end{cases}$

III. Plot the graphs on polar coordinate paper.

1. $r = 3 \sin 3\theta$.
2. $r \cos^2 \frac{1}{2}\theta = 4$.
3. $r = 4 \cos 2\theta$.
4. $r = 4\theta$.
5. $r^2\theta = 9$.

C. An Honor Examination on Algebra.   Time: 2 hours.

1. (a) If r_1 and r_2 are the roots of $ax^2 + bx + c = 0$, what is the value of $r_1^2 r_2 + r_1 r_2^2$?

   (b) Determine k so that there exists only one distinct pair of values satisfying the simultaneous equations:

$$y = 2x + k,$$
$$y = x^2 - 5x + 6.$$

2. (a) Show that by adding one to the product of any four consecutive integers a perfect square is obtained.

   (b) In the expansion of $(1 + x)^{43}$, the numerical coefficients of the $(2r + 1)$th and the $(r + 2)$th term are equal. Find r.

3. (a) For what values of x is $\sqrt{x^2 - 2x - 15}$ real?

   (b) Prove that for any two distinct positive numbers a and b, their geometric mean $(\sqrt{ab})$ is less than their arithmetic mean.

4. (a) Prove:

$$\log\left(\frac{\sqrt{1 + x^2} - 1}{\sqrt{1 + x^2} + 1}\right) = 2[\log(\sqrt{1 + x^2} - 1) - \log x].$$

   (b) Given log 2 = 0.30103. Find the number of zeros between the decimal point and the first significant figure when $(\frac{1}{2})^{1000}$ is expressed as a decimal.

5. In a set of four numbers the first three are in geometric progression, the last three are in arithmetic progression with a common differ-

ence of 6, and the first number is the same as the fourth. Find the numbers.

6. Three men, A, B, and C, set out at the same time to walk a certain distance. A walks $4\frac{1}{2}$ miles an hour and finishes the journey two hours before B, who walks one mile an hour faster than C and finishes the journey in three hours less time. What is the distance?

D. An Examination on Trigonometry. Time: 2 hours.

No tables are needed.

1. Simplify the following expressions by reducing each to a single trigonometric function of the angle x:

(a) $\tan (90° - x)$;

(b) $\sin 3x \cos 2x - \cos 3x \sin 2x$;

(c) $\dfrac{\cos (-x)}{\sin x}$;

(d) $\left[\sin \dfrac{x}{2} + \cos \dfrac{x}{2}\right]^2 - 1$.

2. Given $\sin x = \frac{3}{5}$, $\cos y = -\frac{8}{17}$, and x and y are both in the second quadrant. Find $\cos (x + y)$.

3. Prove the identities:

(a) $\sec x = \tan \left(\dfrac{x}{2} + 45°\right) - \tan x$.

(b) $1 - \sin^2 x = \cos^2 x$.

4. Evaluate:

(a) $\sin (\arctan \frac{5}{12})$.

(b) $\sin x = 2$.

5. Find all values of x between $0°$ and $360°$ satisfying the equation $\cos^2 x - \sin x \cos x = 0$.

6. Given $\triangle ABC$:

(a) with $\alpha = 45°$, $\gamma = 30°$, and $c = 5\sqrt{2}$; find a;

(b) with $a = 3$, $b = 7$, and $\gamma = 120°$; find c;

(c) with $a = 2$, $\alpha = 30°$, and $\gamma = 90°$; find b.

7. State the law of cosines for triangles. Write the formulas for the angles in terms of the sides.

8. Complete the following formulas:

(a) $1 - \cos^2 x = \cdots$;

(b) $\cos^2 x - \sin^2 x = \cdots$;

(c) $\left(\cos \dfrac{x}{2}\right)^2 = \cdots$;

(d) $\cos (x - y) = \cdots$;

(e) $\tan (x + y) = \cdots$;

(f) $\tan \dfrac{x}{2} = \cdots$.

E. **An Examination on Analytic Geometry. Time: 2 hours.**

1. The graph of $y^2 = -8x$ is a conic. Plot this graph and calculate all the parts you know about this conic.
2. Given three points $P_1(4,1)$, $P_2(1,4)$, and $P_3(-2,1)$.

  (a) Find the distances between each pair of points.
  (b) Plot the points.
  (c) Find the equation of the line between P_1 and P_3.
  (d) Find the slopes of the lines P_1P_2 and P_2P_3.
  (e) Find the distance from P_2 to the line P_1P_3.

3. Draw the configuration for the conic $9x^2 - 16y^2 = 144$. Find all appropriate points and equations that exhibit the characteristics of this conic.
4. Transform the equation:

$$5x^2 + 6xy + 5y^2 + 22x - 6y + 21 = 0$$

  into a standard form of a conic. Sketch the curve.
5. A point moves so that the sum of the squares of its distances from the vertices of a triangle is a constant. Prove that its path is a circle.
6. Given $A(-4,9)$, $B(6,5)$, and $C(10,7)$, the coordinates of the vertices of the $\triangle ABC$. Find:

  (a) the coordinates of D, the midpoint of BC;
  (b) the length of AD.
  (c) Is AD perpendicular to BC?

7. Find the equation of the hyperbola with center at the origin, which passes through the point $(-2,3)$, and whose asymptotes have the slopes ± 2.
8. Transform the equation $(x^2 + y^2)^2 = 4(x^2 - y^2)$ to polar coordinates and draw the curve.

Part III — Final Examinations

Exam. #I. Time: 3 hours.

A table of logarithms is needed.

1. (a) Solve for x in $2x^2 - 10x + 17 = 0$.
  (b) Find the sum of the roots.
  (c) Find the product of the roots.

2. Find all the roots of:

$$4x^4 - 34x^3 + 110x^2 - 149x + 51 = 0.$$

3. (a) Expand and simplify $(1 - i)^7$.

(b) Insert two geometric means between 4 and $\frac{32}{27}$.

4. Trace the curve $3x^4 - 11x^3 + 9x^2 - 23 = 0$ and show all its characteristics.

5. Find the center, vertices, foci, eccentricity, semiaxes, equations of the directrices, asymptotes, and draw the curve:

$$9x^2 - 4y^2 = 36.$$

6. Solve the equation: $100 = 25.4(1.02)^x$

by use of logarithms.

7. Given the triangle ABC, in which $a = 1000$, $b = 2340$, and $C = 32° 48'$. Find A, B, and c.

8. Solve the following trigonometric equations for values of x between 0° and 360°:

(a) $\sin \dfrac{x}{2} = \dfrac{1}{2}$.

(b) $\cos 2x - \sin x = 0$.

9. Given the three points $P_1(4,2)$, $P_2(-2,0)$, and $P_3(0,-2)$.

(a) Find the coordinates of the midpoint M of the line joining P_1 and P_3.

(b) Find the equation of the line P_2M.

(c) Find the angle between the lines P_2P_3 and P_1P_3.

10. Sketch the curve $x^2 = 8y$ and find the area between the curve, the X-axis, and the line $x = 4$.

Exam. #II.   Time: 3 hours.

No tables are needed.

1. Complete the following trigonometric formulas:

(a) $\sin (\alpha - \beta) = \cdots$;   (d) $\dfrac{a}{\sin \alpha} = \cdots$;

(b) $\tan (\alpha + \beta) = \cdots$;   (e) $\alpha + \beta + \gamma = \cdots$;

(c) $\sec \theta = \cdots$;   (f) $\log_a MN = \cdots$.

2. Write the following algebraic formulas:

(a) For the solution of a quadratic equation.

(b) For the sum of the first n terms of an A.P.

(c) For the expansion of the binomial $(a + x)^n$.

(d) For the division of a polynomial $f(x)$ by $x - r$.

(e) For the expression of a rational integral equation of the nth degree.

(f) For the permutation of n different things taken r at a time.

3. Write the following plane analytic geometry formulas:
   (a) For the distance between two given points.
   (b) For the midpoint of a line between two given points.
   (c) For the equation of a circle with the center at the origin.
   (d) For the distance from a given point to a given line.
   (e) For the transformation from rectangular coordinates to polar coordinates.
   (f) For the translation of the axes to a new origin.

4. Complete the following calculus formulas:

   (a) $\dfrac{d}{dx}(v^n) = \cdots;$ (d) $\int a\, dv = \cdots;$

   (b) $\dfrac{d}{dx}(uv) = \cdots;$ (e) $\int v^n\, dv = \cdots;$

   (c) $\dfrac{d}{dx}(\cos x) = \cdots;$ (f) $\int \cos v\, dv = \cdots.$

5. Define the following terminology:
   (a) The logarithm of a number.
   (b) An identity.
   (c) An ellipse.
   (d) A quadratic equation in one unknown.
   (e) A variable.
   (f) A ratio.

6. Prove the following:
   (a) $\sin^2 x + \cos^2 x = 1;$
   (b) $\log N^k = k \log N;$
   (c) The Remainder Theorem;
   (d) $d = \sqrt{(x_2 - x_1)^2 + (y_2 - y_1)^2}.$

7. Derive the following formulas:
   (a) The quadratic formula.
   (b) The equation of a circle.
   (c) The area of a triangle in terms of two sides and the included angle.
   (d) The $\sin 2A$ in terms of functions of A.

8. Given two points $P_1(x_1, y_1)$ and $P_2(x_2, y_2)$ at the extremities of a focal chord (latus rectum) of the parabola $y^2 = 4px$. Prove:
   (a) $y_1 y_2 = -4p^2.$
   (b) $x_1 x_2 = p^2.$

Exam. #III. Time: 3 hours.

1. Given the three points $P_1(4,1)$, $P_2(0,5)$, and $P_3(-4,1)$, the vertices of $\triangle P_1P_2P_3$.

   (a) Find the midpoints of the three lines joining these points.
   (b) Find the distance between all pairs of points.
   (c) Find the perimeter of $\triangle P_1P_2P_3$.
   (d) Find the area of $\triangle P_1P_2P_3$.
   (e) Find the equation of P_1P_3.

2. Find all the roots of the equation:
$$2x^4 - 10x^3 + 15x^2 + 10x - 17 = 0.$$

3. Find the center, vertices, foci, eccentricity, semiaxes, equations of the directrices, and asymptotes if any, and draw the curve:
$$9x^2 + 4y^2 = 36.$$

4. (a) Expand and simplify $(1 + i)^6$.
   (b) Insert 2 harmonic means between $\frac{1}{2}$ and 2.

5. Plot the graph of the function:
$$f(x) = x^2 - 7x + 10.$$

Find its maximum or minimum point and zeros.

6. Prove the following trigonometric identities:

   (a) $\sec x \csc x - 2 \cos x \csc x = \tan x - \cot x$.
   (b) $1 - \sin^2 x = \cos^2 x$.
   (c) $\sin 3x = 3 \sin x - 4 \sin^3 x$.

7. (a) Find the center and radius of the circle:
$$x^2 + y^2 - 6x - 8y + 4 = 4.$$

   (b) Find its points of intersection with the coordinate axes.
   (c) Find the equation of the line between the larger x-intercept, P_1, and the larger y-intercept, P_2.
   (d) Prove that the center of the circle lies on the line P_1P_2.
   (e) Find the distance from the origin to the line P_1P_2.

8. A man is hired at a salary of $7800.00 a year and gets paid every two weeks. The following deductions are made:

   $22\frac{1}{2}\%$ for income tax;
   $2\frac{1}{2}\%$ for social security;
   $2.50 for life insurance;
   $3.25 for hospitalization insurance;
   5% for investment in company stock.

What is his "take home" pay check?

What is his annual "take home" pay?

9. Find the area bounded by the curve $y = 6 + x - x^2$ and the X-axis.

10. (a) Find graphically the points of intersection between the line

$y = \dfrac{2}{\pi} x$ and the curve $y = \sin x$.

(b) Find the area bounded by the line and the curve.

ANSWERS TO THE EXERCISES

Exercise I.

1. (a) $5x - 14y$.   (b) 1.

2. (a) $-8a^3b^6c^9$.   (b) $3x^{10}y$.   (c) $\dfrac{x^{5n-4}}{y^{2m+3}}$.

    (d) $\sqrt{a - x}$.   (e) $-5\sqrt{2}$.

3. $4x^4 - 19x^3y + 2x^2y^2 + xy^3 - 6y^4$.

4. $4x^2 - 3xy + 2y^2$.

5. (a) $(3a - b)(3x - 2y)(9x^2 + 6xy + 4y^2)$.

    (b) $(a - 3b)(a + 3b)(a^2 + 9b^2)$.

    (c) $(3x + 2y)(2x + 5y)$.

6. $\dfrac{28y^2 - 12x^2}{(3x + 2y)(2x + 5y)(x - y)}$.

7. 0.

8. $x - 9y$.

9. $\dfrac{13b - 7a - 13\sqrt{a^2 - b^2}}{5a + 13b}$.

10. (a) $2\sqrt{30}$.   (b) $\dfrac{7 - 6i\sqrt{2}}{11}$.

Exercise II.

1.

x	y	Δy	$\Delta y/\Delta x$
-4	-7		
		7	7
-3	0		
		5	5
-2	5		
		3	3
-1	8		
		1	1
0	9		
		-1	-1
1	8		
		-3	-3
2	5		
		-5	-5
3	0		
		-7	-7
4	-7		

3. Zeros: $x = \pm 3$; y-intercept: $y = 9$.

4.

x	0	$\frac{1}{2}$	1	$\frac{3}{2}$	2	$\frac{5}{2}$	3	$\frac{7}{2}$	4
y	-6	$-\frac{15}{8}$	0	$\frac{3}{8}$	0	$-\frac{3}{8}$	0	$\frac{15}{8}$	6

x-intercepts: 1, 2, 3; y-intercept: -6.

5. A straight line; $x = \pm 3$.

6. $2\sqrt{5}$; $\sqrt{34}$; $2\sqrt{5}$.

7. $\frac{\pi}{6}$; $\frac{\pi}{2}$; $\frac{\pi}{36}$; .055268.

8. 45°; 150°; 171.8874°.

9. (a) .9283.　　　(d) .2988.　　　(g) .2397.
　　(b) .8363.　　　(e) .4813.　　　(h) 1.7556.
　　(c) .7618.　　　(f) 4.8430.　　　(i) .7139.

10. (a) 32° 10′.　　(d) 48° 13′.　　(g) 46° 50′.
　　(b) 47° 40′.　　(e) 53° 57′.　　(h) 42° 35′.
　　(c) 41° 30′.　　(f) 46° 45′.　　(i) Impossible.

11. (a) .9511.　　　(c) 1.2423.　　　(e) $-.4036$.
　　(b) $-.9744$.　　(d) -2.9858.　　(f) $-.4436$.

13. $\frac{4}{5}$.

14. 7° 50′.

15.

x	0	$\frac{\pi}{6}$	$\frac{\pi}{4}$	$\frac{\pi}{3}$	$\frac{\pi}{2}$	$\frac{3\pi}{4}$	π	$\frac{5\pi}{4}$	$\frac{3\pi}{2}$	$\frac{7\pi}{4}$	2π
y	0	1.74	2	1.74	0	-2	0	2	0	-2	0

$d = \pi$.

Exercise III.

1. (a) 2.　　　　　(b) $\frac{63}{76}$.　　　　　(c) $-\frac{22}{3}$.
　　(d) 2.　　　　　(e) -2.

4. (a) 19° 13′; 160° 47′.　　　(b) 120°, 180°, 240°.
　　(c) 0°, 45°, 135°, 180°, 225°, 315°.　　　(d) 0°, 270°.
　　(e) 0°, 180°.

7. (a) 60°, 300°.　　　　　(b) 30°, 150°, 270°.
　　(c) 90°, 210°, 270°, 330°.　　(d) 15°, 75°, 195°, 255°.
　　(e) 0°, 60°, 180°, 300°.

8. $4'' \times 8''$.

9. $f = \dfrac{s + rD^2}{D^2 - 1}$; 2.

10. 9.

Exercise IV.

1. 100.

2. (a) $(3x)^{\frac{1}{3}} - \dfrac{2}{3}\dfrac{y}{(3x)^{\frac{2}{3}}} - \dfrac{4}{9}\dfrac{y^2}{(3x)^{\frac{5}{3}}} - \dfrac{40}{81}\dfrac{y^3}{(3x)^{\frac{8}{3}}}.$

   (b) $\dfrac{1}{x} + 4\dfrac{y^{\frac{1}{3}}}{x^{\frac{3}{2}}} + 12\dfrac{y^{\frac{2}{3}}}{x^2} + 32\dfrac{y}{x^{\frac{5}{2}}}.$

   (c) $a^5x^5 + 5a^4bx^4y + 10a^3b^2x^3y^2 + 10a^2b^3x^2y^3.$

3. $\left(\dfrac{77}{\sqrt{2}}\right)\left(\dfrac{3^7}{2^{16}}\right)\dfrac{y^3}{x^{\frac{13}{3}}}.$

4. (a) $x > 2.$ (b) $x > 0.$

5. (a) $1 + i.$ (b) $9 + 19i.$ (c) $3i.$

6. (a) $2(\cos 45° + i \sin 45°).$ (b) $13(\cos 337° 23' + i \sin 337° 23').$

7. (a) $20i.$ (b) $\sqrt{2} + \sqrt{2}i.$

8. (a) $16 + 16\sqrt{3}i.$

   (b) $\sqrt{2}[\cos (75° + k90°) + i \sin (75° + k90°)].$

Exercise V.

2. $(2,3).$

3. (a) $-\frac{3}{2}.$ (b) $4.$ (c) $-\frac{5}{3}.$

4. (a) $y = 3.$ (b) $y + 3x = 8.$ (c) $6y + 5x = 13.$

   (d) $x + 2y = 4.$ (e) $y - 3x + 8 = 0.$ (f) $3y + x + 11 = 0.$

5. $\tan \theta = \frac{1}{7};\quad \theta = 8° 8'.$

6. $\frac{24}{5}.$

7. $x = \frac{24}{13}, y = \frac{10}{13}.$

8. (a) $54.$ (b) $xy_1 + yx_2 + x_1y_2 - yx_1 - xy_2 - x_2y_1.$

   Add $x_1y_1 - x_1y_1$ and prove identity.

9. $\frac{5}{9}, \frac{1}{3}, -\frac{4}{9}.$

10. (a) $-54.$ (b) $-54.$

Exercise VI.

1. (a) $\frac{1}{2}, -\frac{2}{3}.$ (b) $-\frac{7}{4}, -\frac{3}{2}.$ (c) $-\frac{2}{3}.$

2. (a) $\frac{1}{2}(-1 \pm \sqrt{3}i).$ (b) $\frac{1}{2} \pm \sqrt{3}.$ (c) $\frac{3}{4} \pm \frac{1}{2}i.$

4. $x^2 - 9x + 14 = 0.$

5. $2, 7.$

6. (a) $(3,3); (0,-2).$ (b) $(1,-\frac{5}{2}); (-\frac{5}{2},1); (\frac{5}{2},-1); (-1,\frac{5}{2}).$

   (c) $(2,2); (2,-2); (-2,2); (-2,-2).$ (d) $(\pm\frac{1}{5}\sqrt{13}i, \pm\frac{1}{5}\sqrt{41}).$

   (e) $(\frac{4}{11}\sqrt{22}i, \frac{9}{22}\sqrt{22}i); (-\frac{4}{11}\sqrt{22}i, -\frac{9}{22}\sqrt{22}i);$

   $(\frac{1}{3}\sqrt{3}i, -\frac{1}{3}\sqrt{3}i); (-\frac{1}{3}\sqrt{3}i, \frac{1}{3}\sqrt{3}i).$

(f) $(1,1)$; $(-\frac{39}{40} + \frac{3}{40}\sqrt{395}i, -\frac{39}{40} - \frac{3}{40}\sqrt{395}i)$;
$(-\frac{39}{40} - \frac{3}{40}\sqrt{395}i, -\frac{39}{40} + \frac{3}{40}\sqrt{395}i)$.

7. (a) $C(3,7)$; $r = 6$. (b) $C(-\frac{5}{2},0)$; $r = \frac{7}{4}$.

(c) $C(-\frac{1}{3},-\frac{2}{5})$; $r = \dfrac{\sqrt{10}}{3}$.

8. (a) $x^2 + y^2 = 9$. (b) $7x^2 + 7y^2 + 19x - 37y - 222 = 0$.

(c) $x^2 + y^2 - 3x - 5y + 8 = 0$.

9. $x - y = 1$; $(3,2)$, $(-1,-2)$; $4\sqrt{2}$.

10. $(8,1)$.

Exercise VII.

1. (a) $a = 5$, $b = 1$; $e = \dfrac{\sqrt{24}}{5}$; $V_1(5,0)$, $V_2(-5,0)$; $F_1(\sqrt{24},0)$,

$F_2(-\sqrt{24},0)$; L.R. $= \frac{2}{5}$.

(b) $a = \frac{5}{8}$; no semiaxes; $e = 1$; $V(0,0)$; $F(\frac{5}{8},0)$; L.R. $= \frac{5}{2}$.

(c) $a = 4$, $b = 3$; $e = \frac{5}{4}$; $V_1(4,0)$, $V_2(-4,0)$; $F_1(5,0)$, $F_2(-5,0)$;
L.R. $= \frac{9}{2}$.

(d) $a = 5$, $b = 5$; $e = 0$; $V_1(5,0)$, $V_2(-5,0)$.

(e) $a = 1$; no semiaxes; $e = 1$; $V(3,1)$; $F(4,1)$; L.R. $= 4$.

2. (a) $(x - \frac{3}{2})^2 + (y - \frac{5}{2})^2 = 16$; a circle.

(b) $\dfrac{x^2}{4} + \dfrac{y^2}{3} = 1$; an ellipse.

(c) $\dfrac{x'^2}{2} - \dfrac{y'^2}{2} = 1$; a hyperbola.

(d) $\dfrac{(x + 3)^2}{4} + \dfrac{(y - 8)^2}{16} = 1$; an ellipse.

(e) $5x'^2 + 6\sqrt{5}y' = 0$; a parabola.

3. $a = \frac{1}{24}$; $V(\frac{11}{12},\frac{47}{24})$; $F(\frac{11}{12},2)$.

4. $576x^2 + 625y^2 = 360000$.

5. $4x^2 - y^2 = 7$; $y^2 - 4x^2 = 7$.

Exercise VIII.

	θ	0	30°	45°	60°	90°	120°	135°	150°	180°	+
(a)	r	∞	7.46	3.41	2.00	1.00	.67	.59	.54	.50	←
(b)	r	0	.26		.50	.71	.87		.97	1.00	←
(c)	r	∞	9.6	6.4	4.8	3.2	2.4	2.1	1.9	1.6	→
(d)	$\dfrac{r}{a}$	3	2.31	1.41	0	−∞	0	−1.41	−2.31	−3	←
(e)	$\dfrac{r}{a}$	0	.87	0	−.87	0	.87	0	−.87	0	...

$\leftarrow \equiv$ symmetric; $\rightarrow \equiv$ more values needed; . . . $\equiv$ intermediate values needed.

3. (a) $r = 4(2 \sin \theta - \cos \theta)$.

  (b) $r^2(4 \cos^2 \theta - \sin^2 \theta) = 16$.

  (c) $r = \tan^2 \theta \sec \theta$.

4. (a) $3x - 5y + 7 = 0$.

  (b) $4y^2 = 20x + 25$.

  (c) $y^3 + x^2y = 9x$.

5. $r = \dfrac{6}{1 + \cos \theta}$.

Exercise IX.

1. (a) $l = 42$; $s = 297$. (b) $l = \frac{10}{3}$; $s = \frac{77}{6}$.

2. (a) $l = \frac{1}{256}$; $s = \frac{255}{256}$. (b) $l = 27\sqrt{2}$; $s = 13\sqrt{6} + 40\sqrt{2}$.

3. $2, \frac{5}{2}, 3, \frac{7}{2}, 4, \frac{9}{2}, 5, \frac{11}{2}, 6$.

4. $\frac{1}{2}, 2, 8, 32, 128, 512, 2048$.

5. $\frac{19}{55}$.

6. $\frac{2}{11}$.

7. 360.

8. 59,875,200.

9. 924.

10. $C(52,13) \cdot C(39,13) \cdot C(26,13) \cdot C(13,13)$,

  $(635,013,559,600)(8,122,425,444)(10,400,600)$.

11. $\frac{1}{36}, \frac{1}{18}, \frac{1}{12}, \frac{1}{9}, \frac{5}{36}, \frac{1}{6}, \frac{5}{36}, \frac{1}{9}, \frac{1}{12}, \frac{1}{18}, \frac{1}{36}$.

12. 20 cents.

Exercise X.

3. (a) $2, -5, \pm i$. (b) $-1, -\frac{1}{2}, \frac{2}{3}, \frac{3}{4}$. (c) $0, \frac{1}{2}, -2, 3 \pm 2\sqrt{3}$.

5. (a) 1.38. (b) -0.89.

6. (a) $\sqrt[3]{-2} - 2, \omega^2\sqrt[3]{-2} - 2, \omega\sqrt[3]{-2} - 2$.

  (b) $\sqrt[3]{-6} - 2, \omega^2\sqrt[3]{-6} - 2, \omega\sqrt[3]{-6} - 2$.

Exercise XI.

1. $0.5956, 6.6721 - 10, 5.2122$.

2. $0.6827; 222.3$.

3. $9.5186 - 10, 9.5500 - 10$.

4. $59° 14'$.

5. (a) 9.110. (b) 31.05.

6. .8372.

7. (a) $\frac{1}{2}(3 \pm \sqrt{5})$. (b) $\frac{1}{6}(15 \pm \sqrt{345})$.

Exercise XII. (With four-place tables the answers may differ slightly depending upon the method used.)

1. $\beta = 68° 7'$, $b = 34.21$, $c = 36.87$.

2. $\beta = 32° 43'$, $a = 9.958$, $c = 11.84$.

3. $\beta = 49° 50'$, $a = .2399$, $b = .2843$.

4. $\alpha = 48° 24'$, $\beta = 41° 36'$, $c = 151.8$.

5. $\alpha = 28° 45'$, $\beta = 61° 15'$, $b = 82.08$.

6. $b = 610.6$, $c = 451.6$, $\gamma = 43° 50'$, $K = 126600$.

7. $\alpha = 44° 45'$, $\gamma = 73° 03'$, $b = .4894$, $K = .09118$.

8. $\alpha_1 = 62° 17'$, $\beta_1 = 80° 28'$, $b_1 = 440.0$, $K_1 = 52600$.
$\alpha_2 = 117° 43'$, $\beta_2 = 25° 2'$, $b_2 = 188.7$, $K_2 = 22560$.

9. $\alpha = 44° 40'$, $\beta = 49° 36'$, $\gamma = 85° 44'$, $K = 51.70$.

10. $a = \sqrt{7}$, $\beta = 130° 54'$, $\gamma = 19° 6'$, $K = \sqrt{3}$.

11. 9.141 yds.

12. 2263 ft./sec. and 1750 ft./sec.

13. 13°, 266.8 mi./hr., 13° south of west.

Exercise XIII.

2. $P_1P_2 = 7$; $P_1P_3 = \sqrt{293}$; $P_2P_3 = \sqrt{122}$.

3. $-3, -6, -2$; $\dfrac{x-8}{3} = \dfrac{y-3}{6} = \dfrac{z-4}{2}$.

4. $x = 6, y = -3, z = 2$; $x = 4, y = 2, z = 4$;
$x - 2y = 6$, $3z - 2y = 6$, $x + 3z = 6$;
$x + 2y = 4$, $2y + z = 4$, $x + z = 4$;
$\dfrac{1}{\sqrt{14}}(x - 2y + 3z) = \dfrac{6}{\sqrt{14}}$; $\dfrac{1}{\sqrt{6}}(x + 2y + z) = \dfrac{4}{\sqrt{6}}$;
$A_1A_2 + B_1B_2 + C_1C_2 = 0$; $\dfrac{z}{2} = \dfrac{x-5}{-4} = y + \dfrac{1}{2}$.

5. $C(1, -\frac{5}{4}, \frac{3}{2})$; $r = \frac{1}{4}\sqrt{77}$.

Exercise XIV.

1. (a) $e^x[6x \cos 3x^2 + \sin 3x^2] + \dfrac{1}{2\sqrt{x}}\left[7 - \dfrac{1}{\sqrt{1-x}}\right]$.

(b) $6\left[x - 1 + \dfrac{1}{x}\right]$.

(c) $-\dfrac{b^2x}{a^2y}$.

(d) $-\dfrac{2x + D}{2y + E}$.

(e) $1 + (x^{-\frac{1}{2}} + 3x^{\frac{1}{2}})$ arc tan $\sqrt{x}$.

2. (a) $e^x[(3 + 4x - 4x^2) \sin x^2 + (1 + 4x + 4x^2) \cos x^2]$.

   (b) $36x - 2$.

   (c) $-\dfrac{b^4}{a^2 y^3}$.

3. $(-\tfrac{1}{2}, \tfrac{55}{4})$ max; $(3, -72)$ min.

4.

x	y	y'	y''	Remarks
-2	$-\frac{12}{5}$	$-$	$-$	
$-\sqrt{3}$	$-\dfrac{3\sqrt{3}}{2}$	$-$	0	pt. of inf.
-1	-3	0	$+$	min.
0	0	$+$	0	pt. of inf.
1	3	0	$-$	max.
$\sqrt{3}$	$\dfrac{3\sqrt{3}}{2}$	$-$	0	pt. of inf.
2	$\frac{12}{5}$	$-$	$+$	

5. $v = v_1 - 8a_1$; $a = -2a_1$; 600.

6. (a) $x^4 - 2x^3 + 3x + c$.

   (b) $\dfrac{3^x}{\log_e 3} + c$.

   (c) $\tfrac{1}{2}e^{x^2} + c$.

7. (a) 7.   (b) 52.   (c) 2.

8. $\frac{64}{3}$.

Exercise XV.

Answer is shown in Fig. 106.

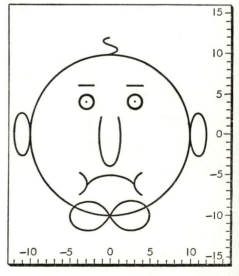

Fig. 106

Examinations — Part I.

Test #1.

1. 2.   **2.** 0.   **3.** x^3.   **4.** -5.   **5.** 1.

Test #2.

1. $F = \frac{9}{5}C + 32°$; $F = 212°$; $C = 35°$.

2. $x = \frac{1}{2}$.

3. 5.

4. (a) $-.3043$.   (b) $.6301$.   (c) $.9652$.

5. $\frac{5}{13}, -\frac{5}{13}$.

Test #3.

1. $\sqrt{3} + 1$.   **4.** 120°.   **5.** $624.

Test #4.

1. $\dfrac{1}{x^{\frac{3}{2}}} + \dfrac{9}{x^2 y} + \dfrac{54}{x^{\frac{5}{2}} y^2} + \dfrac{270}{x^3 y^3} + \dfrac{1215}{x^{\frac{7}{2}} y^4}$.

2. $H = 5\dfrac{x}{y^2}$.

3. $32x^{-18}y^{-10}$.

4. $\sqrt[3]{2}(\cos 15° + i \sin 15°)$; $\sqrt[3]{2}(\cos 135° + i \sin 135°)$; $\sqrt[3]{2}(\cos 255° + i \sin 255°)$.

5. 1.

Test #5.

1. (a) $\sqrt{89}$.   (b) $(1, -\frac{3}{2})$.   (c) $-\frac{5}{8}$.   (d) $8y + 5x + 7 = 0$.
    (e) $5y - 8x = 29$.   (f) $\sqrt{89}$.   (g) $-\frac{13}{3}$.   (h) -1.

2. $u = -\frac{3}{4}, v = \frac{3}{4}, x = \frac{1}{4}$.

Test #6.

1. $\dfrac{1 \pm i}{3}$.   **2.** $(\pm\sqrt{13}i, 2)$; $(\pm\sqrt{7}i, 1)$.

3. $x^2 + y^2 + 2x - 4y = 0$.

4. $y = -x$.

5. $x^2 - 7x + 14 = 0$.

Test #7.

1. $a = \frac{9}{8}$, $V(0,0)$, $F(0,\frac{9}{8})$, Y-axis, $y = -\frac{9}{8}$, L.R. $= \frac{9}{2}$.

2. $36y^2 - 64x^2 = 2304.$
   $V_1(6,0),\ V_2(-6,0),\ V_3(0,8),\ V_4(0,-8).$
   $F_1(10,0),\ F_2(-10,0),\ F_3(0,10),\ F_4(0,-10).$
   $3y = \pm 4x,\ 5x = \pm 18,\ 5y = \pm 32.$

3. $\dfrac{x''^2}{2} + \dfrac{y''^2}{4} = 1,$ an ellipse.

Test #8.

#	θ	0°	30°	60°	90°	120°	150°	180°	210°
1	r	−2.00	.26	2.48	4.00	4.48	3.74	2.00	−.26
2	$\dfrac{r}{a}$	0	.87	.87	0	−.87	−.87	0	.87
5	r	3.00	1.85	1.66	2.00	3.73	−2.5	. . .	. . .

3. $x^2 + y^2 + 2y = 0;$ circle.

4. $r^2(9\cos^2\theta + 16\sin^2\theta) = 144;\ e = \dfrac{\sqrt{7}}{4};$ ellipse.

Test #9.
1. (a) A.P.   (b) $\frac{5}{2}$.   (c) $\frac{21}{2}$.
2. $\frac{127}{495}$.
3. 4,989,600.
4. 252.
5. $.75.

Test #10.
1. $\frac{3}{2},\ -4,\ \pm i.$
3. $3,\ -\frac{1}{2},\ \frac{1}{2}.$
4. Not more than 3 positive and 3 negative roots; ± 1;

x	−3	−2	−1	0	1	2	3
$f(x)$	1001	49	−9	−1	−11	−99	−301

5. 0.14.

Test #11.
1. 4.256.
2. 0.3507.
3. $\frac{1}{2}(5 + \sqrt{33}).$
4. 0.2642.

Test #12.

1. $b = \sqrt{3}$; $c = 2\sqrt{3}$; $A = \frac{3}{2}\sqrt{3}$; $\angle CAB = 60°$.
2. $\alpha = 37° 32'$, $\gamma = 64° 14'$, $b = .6893$ or $b = .6897$.
3. $\frac{5}{4}\sqrt{6479} = 100.6$.
5. $\frac{5}{2}$, $\frac{5}{2}\sqrt{3}$.

Test #13.

1. $5\sqrt{2}$.
2. $c(2, -3, -6)$, $r = 7$.
3. $-2, 6, -2$ or $1, -3, 1$.
4. $x = 2$, $y = -6$, $z = 3$; $3x - y = 6$, $2z - y = 6$, $3x + 2z = 6$.
5. $x - 4 = \dfrac{y - 1}{-3} = z - 3$.

Test #14.

1. (a) $12x^2 - 6x$. (b) $\dfrac{1 - y}{x - 1}$.

2. (a) $\dfrac{x^3}{3} - \dfrac{1}{x} + c$. (b) $-\frac{1}{2}\cos x^2 + c$.

3. $\frac{81}{4}$.

4.

x	y	y'	y''	Remarks
-2	2	$-$	$+$	
-1	-2	0	$+$	min.
0	0	$+$	0	pt. of inf.
1	2	0	$-$	max.
2	-2	$-$	$-$	

5. 96, 32.

Examinations — Part II.

A. 1. (a) $\dfrac{xy}{x^2 + y^2}$. (b) $\dfrac{xy^2 + 3y}{x^2y^2 + 9}$.

2. $z = -\frac{106}{33}$, $(x = \frac{61}{11}, y = \frac{149}{33}, w = -\frac{64}{33})$.

3. (a) $-1 \pm 2i$. (b) $-2, 1$.

4. $(-14, -18)$, $(6, 2)$.

5. (a) 0.5572. (b) $\dfrac{0.3010}{0.4771} = .6309$.

6. $\frac{2720}{9}$.

7. (a) $-\frac{231}{16}x$. **(b)** $-4 + 4i$.

8. (a) $x > 2, x < -\frac{1}{3}$. **(b)** $-\frac{1}{3} < x < 2$.

9. (a) $\frac{7}{55}$. **(b)** $\frac{19}{3}, \frac{32}{3}, 15, \frac{58}{3}, \frac{71}{3}$.

10. $0, -1, \pm\sqrt{3}$.

11. -1.33.

12. (a) 286. **(b)** $\frac{1}{6}, \frac{1}{18}, \frac{2}{9}$.

13. $(5,4)$; $(\frac{1}{2}\sqrt{2}, -\frac{9}{2}\sqrt{2})$.

C. 1. (a) $-\dfrac{bc}{a^2}$. **(b)** $k = -\frac{25}{4}$; $(x = \frac{7}{2}, y = \frac{3}{4})$.

2. (a) $x^4 + 6x^3 + 11x^2 + 6x + 1 = (x^2 + 3x + 1)^2$.
  (b) $r = 14$.

3. (a) $x \le -3, x \ge 5$. **(b)** $\sqrt{ab} < \dfrac{a+b}{2}$.

4. (a) Rationalize the denominator. **(b)** 300.

5. $8, -4, 2, 8$.

6. 18 mi.

D. 1. (a) and (b) $\cot x$, (c) and (d) $\sin x$.

2. $-\frac{13}{85}$.

4. (a) $\frac{5}{13}, 1 - \frac{5}{13}$. **(b)** Impossible.

5. $45°, 90°, 225°, 270°$.

6. (a) 10. **(b)** $\sqrt{79}$. **(c)** $2\sqrt{3}$.

E. 1. $e = 1, V(0,0), F(-2,0), x = 2, \text{L.R.} = 8$.

2. (a) $3\sqrt{2}, 6, 3\sqrt{2}$. **(c)** $y = 1$. **(d)** $-1, 1$. **(e)** 3.

3. $C(0,0)$; $e = \frac{5}{4}$; $V_1(4,0), V_2(-4,0)$; $F_1(5,0), F_2(-5,0)$; $x = \pm\frac{16}{5}$;
  $y = \pm\frac{3}{4}x$; L.R. $= \frac{9}{2}$.

4. $\dfrac{x''^2}{4} + \dfrac{y''^2}{16} = 1$; an ellipse.

5. $x^2 + y^2 + Dx + Ey + F = 0$; $D = -\frac{2}{3}(x_1 + x_2 + x_3)$;
  $E = -\frac{2}{3}(y_1 + y_2 + y_3)$;
  $F = \frac{1}{3}(x_1^2 + x_2^2 + x_3^2 + y_1^2 + y_2^2 + y_3^2 - C)$.

6. (a) $D(8,6)$. **(b)** $\sqrt{153}$. **(c)** No.

7. $4x^2 - y^2 = 7$.

8. $r^2 = 4\cos 2\theta$.

Examinations — Part III.

Exam. #I.

1. (a) $\frac{1}{2}(5 \pm 3i)$. **(b)** 5. **(c)** $\frac{17}{2}$.

2. $\frac{1}{2}, 3, \frac{1}{2}(5 \pm 3i)$.

3. (a) $8 + 8i$. **(b)** $\frac{8}{3}, \frac{16}{9}$.

4. Zeros: -1, $3.-$; min.: $(0,-23)$, $(2,-27)$; max.: $(\frac{3}{4},-21.+)$
pt. of inf.: $(\frac{1}{3},-22.+)$, $(\frac{3}{2},-24.+)$.

5. $a = 2$, $b = 3$; $C(0,0)$; $V_1(2,0)$, $V_2(-2,0)$; $e = \dfrac{\sqrt{13}}{2}$;

$F_1(\sqrt{13},0)$, $F_2(-\sqrt{13},0)$; $x = \pm\frac{4}{13}\sqrt{13}$; $2y = \pm3x$.

6. 69.2.

7. $A = 19°\ 52'$, $B = 127°\ 20'$, $c = 1594$.

8. (a) $60°$, $300°$.　　(b) $30°$, $150°$, $270°$.

9. (a) $M(2,0)$.　　(b) $y = 0$.　　　　(c) $90°$.

10. $\frac{8}{3}$.

Exam. #III

1. (a) $M_{12}(2,3)$, $M_{23}(-2,3)$, $M_{13}(0,1)$.

　(b) $P_1P_2 = 4\sqrt{2}$, $P_1P_3 = 8$, $P_2P_3 = 4\sqrt{2}$.

　(c) $8\sqrt{2} + 8$.

　(d) 16.

　(e) $y = 1$.

2. 1, -1, $\frac{1}{2}(5 \pm 3i)$.

3. $a = 3$, $b = 2$; $C(0,0)$; $V_1(0,3)$, $V_2(0,-3)$; $e = \dfrac{\sqrt{5}}{3}$;

$F_1(0,\sqrt{5})$, $F_2(0,-\sqrt{5})$; $y = \pm\frac{9}{5}\sqrt{5}$.

4. (a) $-8i$.　　　　(b) $\frac{2}{3}$, 1.

5.

x	0	1	2	3	$\frac{7}{2}$	4	5	6
y	10	4	0	-2	$-\frac{9}{4}$	-2	0	4

min. $(\frac{7}{2},-\frac{9}{4})$; zeros at $x = 2$, 5.

7. (a) $C(3,4)$, $r = 5$.　　(b) $(0,0)$, $(6,0)$, $(0,8)$.　　(c) $3y + 4x = 24$.
　(e) $\frac{24}{5}$.

8. \$204.25; \$5310.50.

9. $\frac{125}{6}$.

10. (a) $(0,0)$, $\left(\dfrac{\pi}{2},1\right)$, $\left(-\dfrac{\pi}{2}, -1\right)$.　　(b) $2 - \dfrac{\pi}{2}$.

TABLES

Table I. Powers and Roots.

No.	Square	Cube	Square Root	Cube Root	No.	Square	Cube	Square Root	Cube Root
1	1	1	1.000	1.000	51	2601	132651	7.141	3.708
2	4	8	1.414	1.260	52	2704	140608	7.211	3.733
3	9	27	1.732	1.442	53	2809	148877	7.280	3.756
4	16	64	2.000	1.587	54	2916	157464	7.348	3.780
5	25	125	2.236	1.710	55	3025	166375	7.416	3.803
6	36	216	2.449	1.817	56	3136	175616	7.483	3.826
7	49	343	2.646	1.913	57	3249	185193	7.550	3.849
8	64	512	2.828	2.000	58	3364	195112	7.616	3.871
9	81	729	3.000	2.080	59	3481	205379	7.681	3.893
10	100	1000	3.162	2.154	60	3600	216000	7.746	3.915
11	121	1331	3.317	2.224	61	3721	226981	7.810	3.936
12	144	1728	3.464	2.289	62	3844	238328	7.874	3.958
13	169	2197	3.606	2.351	63	3969	250047	7.937	3.979
14	196	2744	3.742	2.410	64	4096	262144	8.000	4.000
15	225	3375	3.873	2.466	65	4225	274625	8.062	4.021
16	256	4096	4.000	2.520	66	4356	287496	8.124	4.041
17	289	4913	4.123	2.571	67	4489	300763	8.185	4.062
18	324	5832	4.243	2.621	68	4624	314432	8.246	4.082
19	361	6859	4.359	2.668	69	4761	328509	8.307	4.102
20	400	8000	4.472	2.714	70	4900	343000	8.367	4.121
21	441	9261	4.583	2.759	71	5041	357911	8.426	4.141
22	484	10648	4.690	2.802	72	5184	373248	8.485	4.160
23	529	12167	4.796	2.844	73	5329	389017	8.544	4.179
24	576	13824	4.899	2.884	74	5476	405224	8.602	4.198
25	625	15625	5.000	2.924	75	5625	421875	8.660	4.217
26	676	17576	5.099	2.962	76	5776	438976	8.718	4.236
27	729	19683	5.196	3.000	77	5929	456533	8.775	4.254
28	784	21952	5.292	3.037	78	6084	474552	8.832	4.273
29	841	24389	5.385	3.072	79	6241	493039	8.888	4.291
30	900	27000	5.477	3.107	80	6400	512000	8.944	4.309
31	961	29791	5.568	3.141	81	6561	531441	9.000	4.327
32	1024	32768	5.657	3.175	82	6724	551368	9.055	4.344
33	1089	35937	5.745	3.208	83	6889	571787	9.110	4.362
34	1156	39304	5.831	3.240	84	7056	592704	9.165	4.380
35	1225	42875	5.916	3.271	85	7225	614125	9.220	4.397
36	1296	46656	6.000	3.302	86	7396	636056	9.274	4.414
37	1369	50653	6.083	3.332	87	7569	658503	9.327	4.431
38	1444	54872	6.164	3.362	88	7744	681472	9.381	4.448
39	1521	59319	6.245	3.391	89	7921	704969	9.434	4.465
40	1600	64000	6.325	3.420	90	8100	729000	9.487	4.481
41	1681	68921	6.403	3.448	91	8281	753571	9.539	4.498
42	1764	74088	6.481	3.476	92	8464	778688	9.592	4.514
43	1849	79507	6.557	3.503	93	8649	804357	9.644	4.531
44	1936	85184	6.633	3.530	94	8836	830584	9.695	4.547
45	2025	91125	6.708	3.557	95	9025	857375	9.747	4.563
46	2116	97336	6.782	3.583	96	9216	884736	9.798	4.579
47	2209	103823	6.856	3.609	97	9409	912673	9.849	4.595
48	2304	110592	6.928	3.634	98	9604	941192	9.899	4.610
49	2401	117649	7.000	3.659	99	9801	970299	9.950	4.626
50	2500	125000	7.071	3.684	100	10000	1000000	10.000	4.642

Table II. Constants.

Term	Value	Reciprocal	Square	Square Root	Log
$\pi/6$	.52360	1.90986	.27416	.72360	71900
$\pi/4$	.78540	1.27324	.61685	.88623	89509
$\pi/3$	1.04720	0.95493	1.47361	1.02333	02003
$\pi/2$	1.57080	0.63662	2.46740	1.25331	19612
$2\pi/3$	2.09440	0.47746	4.38649	1.44720	32106
$3\pi/4$	2.35619	0.42441	5.55165	1.53499	37221
$5\pi/6$	2.61799	0.38197	6.85389	1.61802	41797
π	3.14159	0.31831	9.86960	1.77245	49715
$7\pi/6$	3.66519	0.27284	13.43363	1.91447	56410
$5\pi/4$	3.92699	0.25465	15.42126	1.98166	59406
$4\pi/3$	4.18879	0.23873	17.54596	2.04665	62209
$3\pi/2$	4.71239	0.21221	22.20661	2.17080	67324
$5\pi/3$	5.23599	0.19099	27.41557	2.28823	71900
$11\pi/6$	5.75959	0.17362	33.17284	2.39991	76039
2π	6.28319	0.15915	39.47842	2.50663	79818
e	2.71828	0.36788	7.38906	1.64872	43429

1 radian = 57.29577 95131 degrees.

1 degree = 0.01745 32925 radians.

Table III.   Four-Place Common Logarithms.

N	0	1	2	3	4	5	6	7	8	9
10	0000	0043	0086	0128	0170	0212	0253	0294	0334	0374
11	0414	0453	0492	0531	0569	0607	0645	0682	0719	0755
12	0792	0828	0864	0899	0934	0969	1004	1038	1072	1106
13	1139	1173	1206	1239	1271	1303	1335	1367	1399	1430
14	1461	1492	1523	1553	1584	1614	1644	1673	1703	1732
15	1761	1790	1818	1847	1875	1903	1931	1959	1987	2014
16	2041	2068	2095	2122	2148	2175	2201	2227	2253	2279
17	2304	2330	2355	2380	2405	2430	2455	2480	2504	2529
18	2553	2577	2601	2625	2648	2672	2695	2718	2742	2765
19	2788	2810	2833	2856	2878	2900	2923	2945	2967	2989
20	3010	3032	3054	3075	3096	3118	3139	3160	3181	3201
21	3222	3243	3263	3284	3304	3324	3345	3365	3385	3404
22	3424	3444	3464	3483	3502	3522	3541	3560	3579	3598
23	3617	3636	3655	3674	3692	3711	3729	3747	3766	3784
24	3802	3820	3838	3856	3874	3892	3909	3927	3945	3962
25	3979	3997	4014	4031	4048	4065	4082	4099	4116	4133
26	4150	4166	4183	4200	4216	4232	4249	4265	4281	4298
27	4314	4330	4346	4362	4378	4393	4409	4425	4440	4456
28	4472	4487	4502	4518	4533	4548	4564	4579	4594	4609
29	4624	4639	4654	4669	4683	4698	4713	4728	4742	4757
30	4771	4786	4800	4814	4829	4843	4857	4871	4886	4900
31	4914	4928	4942	4955	4969	4983	4997	5011	5024	5038
32	5051	5065	5079	5092	5105	5119	5132	5145	5159	5172
33	5185	5198	5211	5224	5237	5250	5263	5276	5289	5302
34	5315	5328	5340	5353	5366	5378	5391	5403	5416	5428
35	5441	5453	5465	5478	5490	5502	5514	5527	5539	5551
36	5563	5575	5587	5599	5611	5623	5635	5647	5658	5670
37	5682	5694	5705	5717	5729	5740	5752	5763	5775	5786
38	5798	5809	5821	5832	5843	5855	5866	5877	5888	5899
39	5911	5922	5933	5944	5955	5966	5977	5988	5999	6010
40	6021	6031	6042	6053	6064	6075	6085	6096	6107	6117
41	6128	6138	6149	6160	6170	6180	6191	6201	6212	6222
42	6232	6243	6253	6263	6274	6284	6294	6304	6314	6325
43	6335	6345	6355	6365	6375	6385	6395	6405	6415	6425
44	6435	6444	6454	6464	6474	6484	6493	6503	6513	6522
45	6532	6542	6551	6561	6571	6580	6590	6599	6609	6618
46	6628	6637	6646	6656	6665	6675	6684	6693	6702	6712
47	6721	6730	6739	6749	6758	6767	6776	6785	6794	6803
48	6812	6821	6830	6839	6848	6857	6866	6875	6884	6893
49	6902	6911	6920	6928	6937	6946	6955	6964	6972	6981
50	6990	6998	7007	7016	7024	7033	7042	7050	7059	7067
51	7076	7084	7093	7101	7110	7118	7126	7135	7143	7152
52	7160	7168	7177	7185	7193	7202	7210	7218	7226	7235
53	7243	7251	7259	7267	7275	7284	7292	7300	7308	7316
54	7324	7332	7340	7348	7356	7364	7372	7380	7388	7396

Table III. Continued.

N	0	1	2	3	4	5	6	7	8	9
55	7404	7412	7419	7427	7435	7443	7451	7459	7466	7474
56	7482	7490	7497	7505	7513	7520	7528	7536	7543	7551
57	7559	7566	7574	7582	7589	7597	7604	7612	7619	7627
58	7634	7642	7649	7657	7664	7672	7679	7686	7694	7701
59	7709	7716	7723	7731	7738	7745	7752	7760	7767	7774
60	7782	7789	7796	7803	7810	7818	7825	7832	7839	7846
61	7853	7860	7868	7875	7882	7889	7896	7903	7910	7917
62	7924	7931	7938	7945	7952	7959	7966	7973	7980	7987
63	7993	8000	8007	8014	8021	8028	8035	8041	8048	8055
64	8062	8069	8075	8082	8089	8096	8102	8109	8116	8122
65	8129	8136	8142	8149	8156	8162	8169	8176	8182	8189
66	8195	8202	8209	8215	8222	8228	8235	8241	8248	8254
67	8261	8267	8274	8280	8287	8293	8299	8306	8312	8319
68	8325	8331	8338	8344	8351	8357	8363	8370	8376	8382
69	8388	8395	8401	8407	8414	8420	8426	8432	8439	8445
70	8451	8457	8463	8470	8476	8482	8488	8494	8500	8506
71	8513	8519	8525	8531	8537	8543	8549	8555	8561	8567
72	8573	8579	8585	8591	8597	8603	8609	8615	8621	8627
73	8633	8639	8645	8651	8657	8663	8669	8675	8681	8686
74	8692	8698	8704	8710	8716	8722	8727	8733	8739	8745
75	8751	8756	8762	8768	8774	8779	8785	8791	8797	8802
76	8808	8814	8820	8825	8831	8837	8842	8848	8854	8859
77	8865	8871	8876	8882	8887	8893	8899	8904	8910	8915
78	8921	8927	8932	8938	8943	8949	8954	8960	8965	8971
79	8976	8982	8987	8993	8998	9004	9009	9015	9020	9025
80	9031	9036	9042	9047	9053	9058	9063	9069	9074	9079
81	9085	9090	9096	9101	9106	9112	9117	9122	9128	9133
82	9138	9143	9149	9154	9159	9165	9170	9175	9180	9186
83	9191	9196	9201	9206	9212	9217	9222	9227	9232	9238
84	9243	9248	9253	9258	9263	9269	9274	9279	9284	9289
85	9294	9299	9304	9309	9315	9320	9325	9330	9335	9340
86	9345	9350	9355	9360	9365	9370	9375	9380	9385	9390
87	9395	9400	9405	9410	9415	9420	9425	9430	9435	9440
88	9445	9450	9455	9460	9465	9469	9474	9479	9484	9489
89	9494	9499	9504	9509	9513	9518	9523	9528	9533	9538
90	9542	9547	9552	9557	9562	9566	9571	9576	9581	9586
91	9590	9595	9600	9605	9609	9614	9619	9624	9628	9633
92	9638	9643	9647	9652	9657	9661	9666	9671	9675	9680
93	9685	9689	9694	9699	9703	9708	9713	9717	9722	9727
94	9731	9736	9741	9745	9750	9754	9759	9763	9768	9773
95	9777	9782	9786	9791	9795	9800	9805	9809	9814	9818
96	9823	9827	9832	9836	9841	9845	9850	9854	9859	9863
97	9868	9872	9877	9881	9886	9890	9894	9899	9903	9908
98	9912	9917	9921	9926	9930	9934	9939	9943	9948	9952
99	9956	9961	9965	9969	9974	9978	9983	9987	9991	9996

Table IV. Logarithms of Trigonometric Functions

Angle		L sin	d	L cos	d	L tan	cd	L cot	Angle	
0°	0'	- ∞		0.0000	0	- ∞		∞	90°	0'
	10'	7.4637	3011	0.0000	0	7.4637	3011	2.5363		50'
	20'	7.7648	1760	0.0000	0	7.7648	1761	2.2352		40'
	30'	7.9408	1250	0.0000	0	7.9409	1249	2.0591		30'
	40'	8.0658	969	0.0000	0	8.0658	969	1.9342		20'
	50'	8.1627	792	0.0000	1	8.1627	792	1.8373		10'
1°	0'	8.2419	669	9.9999	0	8.2419	670	1.7581	89°	0'
	10'	8.3088	580	9.9999	0	8.3089	580	1.6911		50'
	20'	8.3668	511	9.9999	0	8.3669	512	1.6331		40'
	30'	8.4179	458	9.9999	1	8.4181	457	1.5819		30'
	40'	8.4637	413	9.9998	0	8.4638	415	1.5362		20'
	50'	8.5050	378	9.9998	1	8.5053	378	1.4947		10'
2°	0'	8.5428	348	9.9997	0	8.5431	348	1.4569	88°	0'
	10'	8.5776	321	9.9997	1	8.5779	322	1.4221		50'
	20'	8.6097	300	9.9996	0	8.6101	300	1.3899		40'
	30'	8.6397	280	9.9996	1	8.6401	281	1.3599		30'
	40'	8.6677	263	9.9995	0	8.6682	263	1.3318		20'
	50'	8.6940	248	9.9995	1	8.6945	249	1.3055		10'
3°	0'	8.7188	235	9.9994	1	8.7194	235	1.2806	87°	0'
	10'	8.7423	222	9.9993	0	8.7429	223	1.2571		50'
	20'	8.7645	212	9.9993	1	8.7652	213	1.2348		40'
	30'	8.7857	202	9.9992	1	8.7865	202	1.2135		30'
	40'	8.8059	192	9.9991	1	8.8067	194	1.1933		20'
	50'	8.8251	185	9.9990	1	8.8261	185	1.1739		10'
4°	0'	8.8436	177	9.9989	1	8.8446	178	1.1554	86°	0'
	10'	8.8613	170	9.9988	0	8.8624	171	1.1376		50'
	20'	8.8783	163	9.9988	1	8.8795	165	1.1205		40'
	30'	8.8946	158	9.9987	1	8.8960	158	1.1040		30'
	40'	8.9104	152	9.9986	2	8.9118	154	1.0882		20'
	50'	8.9256	147	9.9984	1	8.9272	148	1.0728		10'
5°	0'	8.9403	142	9.9983	1	8.9420	143	1.0580	85°	0'
	10'	8.9545	137	9.9982	1	8.9563	138	1.0437		50'
	20'	8.9682	134	9.9981	1	8.9701	135	1.0299		40'
	30'	8.9816	129	9.9980	1	8.9836	130	1.0164		30'
	40'	8.9945	125	9.9979	2	8.9966	127	1.0034		20'
	50'	9.0070	122	9.9977	1	9.0093	123	0.9907		10'
6°	0'	9.0192	119	9.9976	1	9.0216	120	0.9784	84°	0'
	10'	9.0311	115	9.9975	2	9.0336	117	0.9664		50'
	20'	9.0426	113	9.9973	1	9.0453	114	0.9547		40'
	30'	9.0539	109	9.9972	2	9.0567	111	0.9433		30'
	40'	9.0648	107	9.9970	1	9.0678	108	0.9322		20'
	50'	9.0755	104	9.9969	1	9.0786	105	0.9214		10'
7°	0'	9.0859	102	9.9968	2	9.0891	104	0.9109	83°	0'
	10'	9.0961	99	9.9966	2	9.0995	101	0.9005		50'
	20'	9.1060	97	9.9964	1	9.1096	98	0.8904		40'
	30'	9.1157	95	9.9963	2	9.1194	97	0.8806		30'
Angle		L cos	d	L sin	d	L cot	cd	L tan	Angle	

Table IV. Continued.

Angle	L sin	d	L cos	d	L tan	cd	L cot	Angle
7° 30'	9.1157	95	9.9963	2	9.1194	97	0.8806	83° 30'
40'	9.1252	93	9.9961	2	9.1291	94	0.8709	20'
50'	9.1345	91	9.9959	1	9.1385	93	0.8615	10'
8° 0'	9.1436	89	9.9958	2	9.1478	91	0.8522	82° 0'
10'	9.1525	87	9.9956	2	9.1569	89	0.8431	50'
20'	9.1612	85	9.9954	2	9.1658	87	0.8342	40'
30'	9.1697	84	9.9952	2	9.1745	86	0.8255	30'
40'	9.1781	82	9.9950	2	9.1831	84	0.8169	20'
50'	9.1863	80	9.9948	2	9.1915	82	0.8085	10'
9° 0'	9.1943	79	9.9946	2	9.1997	81	0.8003	81° 0'
10'	9.2022	78	9.9944	2	9.2078	80	0.7922	50'
20'	9.2100	76	9.9942	2	9.2158	78	0.7842	40'
30'	9.2176	75	9.9940	2	9.2236	77	0.7764	30'
40'	9.2251	73	9.9938	2	9.2313	76	0.7687	20'
50'	9.2324	73	9.9936	2	9.2389	74	0.7611	10'
10° 0'	9.2397	71	9.9934	3	9.2463	73	0.7537	80° 0'
10'	9.2468	70	9.9931	2	9.2536	73	0.7464	50'
20'	9.2538	68	9.9929	2	9.2609	71	0.7391	40'
30'	9.2606	68	9.9927	3	9.2680	70	0.7320	30'
40'	9.2674	66	9.9924	2	9.2750	69	0.7250	20'
50'	9.2740	66	9.9922	2	9.2819	68	0.7181	10'
11° 0'	9.2806	64	9.9919	2	9.2887	66	0.7113	79° 0'
10'	9.2870	64	9.9917	3	9.2953	67	0.7047	50'
20'	9.2934	63	9.9914	2	9.3020	65	0.6980	40'
30'	9.2997	61	9.9912	3	9.3085	64	0.6915	30'
40'	9.3058	61	9.9909	3	9.3149	63	0.6851	20'
50'	9.3119	60	9.9907	3	9.3212	63	0.6788	10'
12° 0'	9.3179	59	9.9904	3	9.3275	61	0.6725	78° 0'
10'	9.3238	58	9.9901	2	9.3336	61	0.6664	50'
20'	9.3296	57	9.9899	3	9.3397	61	0.6603	40'
30'	9.3353	57	9.9896	3	9.3458	59	0.6542	30'
40'	9.3410	56	9.9893	3	9.3517	59	0.6483	20'
50'	9.3466	55	9.9890	3	9.3576	58	0.6424	10'
13° 0'	9.3521	54	9.9887	3	9.3634	57	0.6366	77° 0'
10'	9.3575	54	9.9884	3	9.3691	57	0.6309	50'
20'	9.3629	53	9.9881	3	9.3748	56	0.6252	40'
30'	9.3682	52	9.9878	3	9.3804	55	0.6196	30'
40'	9.3734	52	9.9875	3	9.3859	55	0.6141	20'
50'	9.3786	51	9.9872	3	9.3914	54	0.6086	10'
14° 0'	9.3837	50	9.9869	3	9.3968	53	0.6032	76° 0'
10'	9.3887	50	9.9866	3	9.4021	53	0.5979	50'
20'	9.3937	49	9.9863	4	9.4074	53	0.5926	40'
30'	9.3986	49	9.9859	3	9.4127	51	0.5873	30'
40'	9.4035	48	9.9856	3	9.4178	52	0.5822	20'
50'	9.4083	47	9.9853	4	9.4230	50	0.5770	10'
Angle	L cos	d	L sin	d	L cot	cd	L tan	Angle

Table IV.   Continued.

Angle	L sin	d	L cos	d	L tan	cd	L cot	Angle
15° 0'	9.4130	47	9.9849	3	9.4280	51	0.5720	75° 0'
10'	9.4177	46	9.9846	3	9.4331	50	0.5669	50'
20'	9.4223	46	9.9843	4	9.4381	49	0.5619	40'
30'	9.4269	45	9.9839	3	9.4430	49	0.5570	30'
40'	9.4314	45	9.9836	4	9.4479	48	0.5521	20'
50'	9.4359	44	9.9832	4	9.4527	48	0.5473	10'
16° 0'	9.4403	44	9.9828	3	9.4575	47	0.5425	74° 0'
10'	9.4447	44	9.9825	4	9.4622	47	0.5378	50'
20'	9.4491	42	9.9821	4	9.4669	47	0.5331	40'
30'	9.4533	43	9.9817	3	9.4716	46	0.5284	30'
40'	9.4576	42	9.9814	4	9.4762	46	0.5238	20'
50'	9.4618	41	9.9810	4	9.4808	45	0.5192	10'
17° 0'	9.4659	41	9.9806	4	9.4853	45	0.5147	73° 0'
10'	9.4700	41	9.9802	4	9.4898	45	0.5102	50'
20'	9.4741	40	9.9798	4	9.4943	44	0.5057	40'
30'	9.4781	40	9.9794	4	9.4987	44	0.5013	30'
40'	9.4821	40	9.9790	4	9.5031	44	0.4969	20'
50'	9.4861	39	9.9786	4	9.5075	43	0.4925	10'
18° 0'	9.4900	39	9.9782	4	9.5118	43	0.4882	72° 0'
10'	9.4939	38	9.9778	4	9.5161	42	0.4839	50'
20'	9.4977	38	9.9774	4	9.5203	42	0.4797	40'
30'	9.5015	37	9.9770	5	9.5245	42	0.4755	30'
40'	9.5052	38	9.9765	4	9.5287	42	0.4713	20'
50'	9.5090	36	9.9761	4	9.5329	41	0.4671	10'
19° 0'	9.5126	37	9.9757	5	9.5370	41	0.4630	71° 0'
10'	9.5163	36	9.9752	4	9.5411	40	0.4589	50'
20'	9.5199	36	9.9748	5	9.5451	40	0.4549	40'
30'	9.5235	35	9.9743	4	9.5491	40	0.4509	30'
40'	9.5270	36	9.9739	5	9.5531	40	0.4469	20'
50'	9.5306	34	9.9734	4	9.5571	40	0.4429	10'
20° 0'	9.5340	35	9.9730	5	9.5611	39	0.4389	70° 0'
10'	9.5375	34	9.9725	4	9.5650	39	0.4350	50'
20'	9.5409	34	9.9721	5	9.5689	38	0.4311	40'
30'	9.5443	34	9.9716	5	9.5727	39	0.4273	30'
40'	9.5477	33	9.9711	5	9.5766	38	0.4234	20'
50'	9.5510	33	9.9706	4	9.5804	38	0.4196	10'
21° 0'	9.5543	33	9.9702	5	9.5842	37	0.4158	69° 0'
10'	9.5576	33	9.9697	5	9.5879	38	0.4121	50'
20'	9.5609	32	9.9692	5	9.5917	37	0.4083	40'
30'	9.5641	32	9.9687	5	9.5954	37	0.4046	30'
40'	9.5673	31	9.9682	5	9.5991	37	0.4009	20'
50'	9.5704	32	9.9677	5	9.6028	36	0.3972	10'
22° 0'	9.5736	31	9.9672	5	9.6064	36	0.3936	68° 0'
10'	9.5767	31	9.9667	6	9.6100	36	0.3900	50'
20'	9.5798	30	9.9661	5	9.6136	36	0.3864	40'
30'	9.5828	31	9.9656	5	9.6172	36	0.3828	30'
Angle	L cos	d	L sin	d	L cot	cd	L tan	Angle

Table IV. Continued.

Angle	L sin	d	L cos	d	L tan	cd	L cot	Angle
22° 30'	9.5828	31	9.9656	5	9.6172	36	0.3828	68° 30'
40'	9.5859	30	9.9651	5	9.6208	35	0.3792	20'
50'	9.5889	30	9.9646	6	9.6243	36	0.3757	10'
23° 0'	9.5919	29	9.9640	5	9.6279	35	0.3721	67° 0'
10'	9.5948	30	9.9635	6	9.6314	34	0.3686	50'
20'	9.5978	29	9.9629	5	9.6348	35	0.3652	40'
30'	9.6007	29	9.9624	6	9.6383	34	0.3617	30'
40'	9.6036	29	9.9618	5	9.6417	35	0.3583	20'
50'	9.6065	28	9.9613	6	9.6452	34	0.3548	10'
24° 0'	9.6093	28	9.9607	5	9.6486	34	0.3514	66° 0'
10'	9.6121	28	9.9602	6	9.6520	33	0.3480	50'
20'	9.6149	28	9.9596	6	9.6553	34	0.3447	40'
30'	9.6177	28	9.9590	6	9.6587	33	0.3413	30'
40'	9.6205	27	9.9584	5	9.6620	34	0.3380	20'
50'	9.6232	27	9.9579	6	9.6654	33	0.3346	10'
25° 0'	9.6259	27	9.9573	6	9.6687	33	0.3313	65° 0'
10'	9.6286	27	9.9567	6	9.6720	32	0.3280	50'
20'	9.6313	27	9.9561	6	9.6752	33	0.3248	40'
30'	9.6340	26	9.9555	6	9.6785	32	0.3215	30'
40'	9.6366	26	9.9549	6	9.6817	33	0.3183	20'
50'	9.6392	26	9.9543	6	9.6850	32	0.3150	10'
26° 0'	9.6418	26	9.9537	7	9.6882	32	0.3118	64° 0'
10'	9.6444	26	9.9530	6	9.6914	32	0.3086	50'
20'	9.6470	25	9.9524	6	9.6946	31	0.3054	40'
30'	9.6495	26	9.9518	6	9.6977	32	0.3023	30'
40'	9.6521	25	9.9512	7	9.7009	31	0.2991	20'
50'	9.6546	24	9.9505	6	9.7040	32	0.2960	10'
27° 0'	9.6570	25	9.9499	7	9.7072	31	0.2928	63° 0'
10'	9.6595	25	9.9492	6	9.7103	31	0.2897	50'
20'	9.6620	24	9.9486	7	9.7134	31	0.2866	40'
30'	9.6644	24	9.9479	6	9.7165	31	0.2835	30'
40'	9.6668	24	9.9473	7	9.7196	30	0.2804	20'
50'	9.6692	24	9.9466	7	9.7226	31	0.2774	10'
28° 0'	9.6716	24	9.9459	6	9.7257	30	0.2743	62° 0'
10'	9.6740	23	9.9453	7	9.7287	30	0.2713	50'
20'	9.6763	24	9.9446	7	9.7317	31	0.2683	40'
30'	9.6787	23	9.9439	7	9.7348	30	0.2652	30'
40'	9.6810	23	9.9432	7	9.7378	30	0.2622	20'
50'	9.6833	23	9.9425	7	9.7408	30	0.2592	10'
29° 0'	9.6856	22	9.9418	7	9.7438	29	0.2562	61° 0'
10'	9.6878	23	9.9411	7	9.7467	30	0.2533	50'
20'	9.6901	22	9.9404	7	9.7497	29	0.2503	40'
30'	9.6923	23	9.9397	7	9.7526	30	0.2474	30'
40'	9.6946	22	9.9390	7	9.7556	29	0.2444	20'
50'	9.6968	22	9.9383	8	9.7585	29	0.2415	10'
Angle	L cos	d	L sin	d	L cot	cd	L tan	Angle

Table IV. Continued.

Angle		L sin	d	L cos	d	L tan	cd	L cot	Angle	
30°	0'	9.6990	22	9.9375	7	9.7614	30	0.2386	60°	0'
	10'	9.7012	21	9.9368	7	9.7644	30	0.2356		50'
	20'	9.7033	22	9.9361	7	9.7673	29	0.2327		40'
	30'	9.7055	21	9.9353	8	9.7701	28	0.2299		30'
	40'	9.7076	21	9.9346	7	9.7730	29	0.2270		20'
	50'	9.7097	21	9.9338	8	9.7759	29	0.2241		10'
			21		7		29			
31°	0'	9.7118	21	9.9331	8	9.7788	28	0.2212	59°	0'
	10'	9.7139	21	9.9323	8	9.7816	29	0.2184		50'
	20'	9.7160	21	9.9315	7	9.7845	28	0.2155		40'
	30'	9.7181	20	9.9308	8	9.7873	29	0.2127		30'
	40'	9.7201	21	9.9300	8	9.7902	28	0.2098		20'
	50'	9.7222	20	9.9292	8	9.7930	28	0.2070		10'
32°	0'	9.7242	20	9.9284	8	9.7958	28	0.2042	58°	0'
	10'	9.7262	20	9.9276	8	9.7986	28	0.2014		50'
	20'	9.7282	20	9.9268	8	9.8014	28	0.1986		40'
	30'	9.7302	20	9.9260	8	9.8042	28	0.1958		30'
	40'	9.7322	20	9.9252	8	9.8070	27	0.1930		20'
	50'	9.7342	19	9.9244	8	9.8097	28	0.1903		10'
33°	0'	9.7361	19	9.9236	8	9.8125	28	0.1875	57°	0'
	10'	9.7380	20	9.9228	9	9.8153	27	0.1847		50'
	20'	9.7400	19	9.9219	8	9.8180	28	0.1820		40'
	30'	9.7419	19	9.9211	8	9.8208	27	0.1792		30'
	40'	9.7438	19	9.9203	9	9.8235	28	0.1765		20'
	50'	9.7457	19	9.9194	8	9.8263	27	0.1737		10'
34°	0'	9.7476	18	9.9186	9	9.8290	27	0.1710	56°	0'
	10'	9.7494	19	9.9177	8	9.8317	27	0.1683		50'
	20'	9.7513	18	9.9169	9	9.8344	27	0.1656		40'
	30'	9.7531	19	9.9160	9	9.8371	27	0.1629		30'
	40'	9.7550	18	9.9151	9	9.8398	27	0.1602		20'
	50'	9.7568	18	9.9142	8	9.8425	27	0.1575		10'
35°	0'	9.7586	18	9.9134	9	9.8452	27	0.1548	55°	0'
	10'	9.7604	18	9.9125	9	9.8479	27	0.1521		50'
	20'	9.7622	18	9.9116	9	9.8506	27	0.1494		40'
	30'	9.7640	17	9.9107	9	9.8533	26	0.1467		30'
	40'	9.7657	18	9.9098	9	9.8559	27	0.1441		20'
	50'	9.7675	17	9.9089	9	9.8586	27	0.1414		10'
36°	0'	9.7692	18	9.9080	10	9.8613	26	0.1387	54°	0'
	10'	9.7710	17	9.9070	9	9.8639	27	0.1361		50'
	20'	9.7727	17	9.9061	9	9.8666	26	0.1334		40'
	30'	9.7744	17	9.9052	10	9.8692	26	0.1308		30'
	40'	9.7761	17	9.9042	9	9.8718	27	0.1282		20'
	50'	9.7778	17	9.9033	10	9.8745	26	0.1255		10'
37°	0'	9.7795	16	9.9023	9	9.8771	26	0.1229	53°	0'
	10'	9.7811	17	9.9014	10	9.8797	27	0.1203		50'
	20'	9.7828	16	9.9004	9	9.8824	26	0.1176		40'
	30'	9.7844	17	9.8995	10	9.8850	26	0.1150		30'
Angle		L cos	d	L sin	d	L cot	cd	L tan	Angle	

Table IV. Continued.

Angle		L sin	d	L cos	d	L tan	cd	L cot	Angle	
37°	40'	9.7861	16	9.8985	10	9.8876	26	0.1124	53°	20'
	50'	9.7877	16	9.8975	10	9.8902	26	0.1098		10'
38°	0'	9.7893	17	9.8965	10	9.8928	26	0.1072	52°	0'
	10'	9.7910	16	9.8955	10	9.8954	26	0.1046		50'
	20'	9.7926	15	9.8945	10	9.8980	26	0.1020		40'
	30'	9.7941	16	9.8935	10	9.9006	26	0.0994		30'
	40'	9.7957	16	9.8925	10	9.9032	26	0.0968		20'
	50'	9.7973	16	9.8915	10	9.9058	26	0.0942		10'
39°	0'	9.7989	15	9.8905	10	9.9084	26	0.0916	51°	0'
	10'	9.8004	16	9.8895	11	9.9110	25	0.0890		50'
	20'	9.8020	15	9.8884	10	9.9135	26	0.0865		40'
	30'	9.8035	15	9.8874	10	9.9161	26	0.0839		30'
	40'	9.8050	16	9.8864	11	9.9187	25	0.0813		20'
	50'	9.8066	15	9.8853	10	9.9212	26	0.0788		10'
40°	0'	9.8081	15	9.8843	11	9.9238	26	0.0762	50°	0'
	10'	9.8096	15	9.8832	11	9.9264	25	0.0736		50'
	20'	9.8111	14	9.8821	11	9.9289	26	0.0711		40'
	30'	9.8125	15	9.8810	10	9.9315	26	0.0685		30'
	40'	9.8140	15	9.8800	11	9.9341	25	0.0659		20'
	50'	9.8155	14	9.8789	11	9.9366	26	0.0634		10'
41°	0'	9.8169	15	9.8778	11	9.9392	25	0.0608	49°	0'
	10'	9.8184	14	9.8767	11	9.9417	26	0.0583		50'
	20'	9.8198	15	9.8756	11	9.9443	25	0.0557		40'
	30'	9.8213	14	9.8745	12	9.9468	26	0.0532		30'
	40'	9.8227	14	9.8733	11	9.9494	25	0.0506		20'
	50'	9.8241	14	9.8722	11	9.9519	25	0.0481		10'
42°	0'	9.8255	14	9.8711	12	9.9544	26	0.0456	48°	0'
	10'	9.8269	14	9.8699	11	9.9570	25	0.0430		50'
	20'	9.8283	14	9.8688	12	9.9595	26	0.0405		40'
	30'	9.8297	14	9.8676	11	9.9621	25	0.0379		30'
	40'	9.8311	13	9.8665	12	9.9646	25	0.0354		20'
	50'	9.8324	14	9.8653	12	9.9671	26	0.0329		10'
43°	0'	9.8338	13	9.8641	12	9.9697	25	0.0303	47°	0'
	10'	9.8351	14	9.8629	11	9.9722	25	0.0278		50'
	20'	9.8365	13	9.8618	12	9.9747	25	0.0253		40'
	30'	9.8378	13	9.8606	12	9.9772	26	0.0228		30'
	40'	9.8391	14	9.8594	12	9.9798	25	0.0202		20'
	50'	9.8405	13	9.8582	13	9.9823	25	0.0177		10'
44°	0'	9.8418	13	9.8569	12	9.9848	26	0.0152	46°	0'
	10'	9.8431	13	9.8557	12	9.9874	25	0.0126		50'
	20'	9.8444	13	9.8545	13	9.9899	25	0.0101		40'
	30'	9.8457	12	9.8532	12	9.9924	25	0.0076		30'
	40'	9.8469	13	9.8520	13	9.9949	26	0.0051		20'
	50'	9.8482	13	9.8507	12	9.9975	25	0.0025		10'
45°	0'	9.8495		9.8495		0.0000		0.0000	45°	0'
Angle		L cos	d	L sin	d	L cot	cd	L tan	Angle	

Table V. Natural Trigonometric Functions.

Angle		sin	cos	tan	cot	Angle	
0°	0'	.0000	1.0000	.0000	∞	90°	0'
	10'	.0029	1.0000	.0029	343.774		50'
	20'	.0058	1.0000	.0058	171.885		40'
	30'	.0087	1.0000	.0087	114.589		30'
	40'	.0116	.9999	.0116	85.9398		20'
	50'	.0145	.9999	.0145	68.7501		10'
1°	0'	.0175	.9998	.0175	57.2900	89°	0'
	10'	.0204	.9998	.0204	49.1039		50'
	20'	.0233	.9997	.0233	42.9641		40'
	30'	.0262	.9997	.0262	38.1885		30'
	40'	.0291	.9996	.0291	34.3678		20'
	50'	.0320	.9995	.0320	31.2416		10'
2°	0'	.0349	.9994	.0349	28.6363	88°	0'
	10'	.0378	.9993	.0378	26.4316		50'
	20'	.0407	.9992	.0407	24.5418		40'
	30'	.0436	.9990	.0437	22.9038		30'
	40'	.0465	.9989	.0466	21.4704		20'
	50'	.0494	.9988	.0495	20.2056		10'
3°	0'	.0523	.9986	.0524	19.0811	87°	0'
	10'	.0552	.9985	.0553	18.0750		50'
	20'	.0581	.9983	.0582	17.1693		40'
	30'	.0610	.9981	.0612	16.3499		30'
	40'	.0640	.9980	.0641	15.6048		20'
	50'	.0669	.9978	.0670	14.9244		10'
4°	0'	.0698	.9976	.0699	14.3007	86°	0'
	10'	.0727	.9974	.0729	13.7267		50'
	20'	.0756	.9971	.0758	13.1969		40'
	30'	.0785	.9969	.0787	12.7062		30'
	40'	.0814	.9967	.0816	12.2505		20'
	50'	.0843	.9964	.0846	11.8262		10'
5°	0'	.0872	.9962	.0875	11.4301	85°	0'
	10'	.0901	.9959	.0904	11.0594		50'
	20'	.0929	.9957	.0934	10.7119		40'
	30'	.0958	.9954	.0963	10.3854		30'
	40'	.0987	.9951	.0992	10.0780		20'
	50'	.1016	.9948	.1022	9.7882		10'
6°	0'	.1045	.9945	.1051	9.5144	84°	0'
	10'	.1074	.9942	.1080	9.2553		50'
	20'	.1103	.9939	.1110	9.0098		40'
	30'	.1132	.9936	.1139	8.7769		30'
	40'	.1161	.9932	.1169	8.5555		20'
	50'	.1190	.9929	.1198	8.3450		10'
7°	0'	.1219	.9925	.1228	8.1443	83°	0'
	10'	.1248	.9922	.1257	7.9530		50'
	20'	.1276	.9918	.1287	7.7704		40'
	30'	.1305	.9914	.1317	7.5958		30'
Angle		cos	sin	cot	tan	Angle	

Table V.   Continued.

Angle		sin	cos	tan	cot	Angle	
	30'	.1305	.9914	.1317	7.5958		30'
	40'	.1334	.9911	.1346	7.4287		20'
	50'	.1363	.9907	.1376	7.2687		10'
8°	0'	.1392	.9903	.1405	7.1154	82°	0'
	10'	.1421	.9899	.1435	6.9682		50'
	20'	.1449	.9894	.1465	6.8269		40'
	30'	.1478	.9890	.1495	6.6912		30'
	40'	.1507	.9886	.1524	6.5606		20'
	50'	.1536	.9881	.1554	6.4348		10'
9°	0'	.1564	.9877	.1584	6.3138	81°	0'
	10'	.1593	.9872	.1614	6.1970		50'
	20'	.1622	.9868	.1644	6.0844		40'
	30'	.1650	.9863	.1673	5.9758		30'
	40'	.1679	.9858	.1703	5.8708		20'
	50'	.1708	.9853	.1733	5.7694		10'
10°	0'	.1736	.9848	.1763	5.6713	80°	0'
	10'	.1765	.9843	.1793	5.5764		50'
	20'	.1794	.9838	.1823	5.4845		40'
	30'	.1822	.9833	.1853	5.3955		30'
	40'	.1851	.9827	.1883	5.3093		20'
	50'	.1880	.9822	.1914	5.2257		10'
11°	0'	.1908	.9816	.1944	5.1446	79°	0'
	10'	.1937	.9811	.1974	5.0658		50'
	20'	.1965	.9805	.2004	4.9894		40'
	30'	.1994	.9799	.2035	4.9152		30'
	40'	.2022	.9793	.2065	4.8430		20'
	50'	.2051	.9787	.2095	4.7729		10'
12°	0'	.2079	.9781	.2126	4.7046	78°	0'
	10'	.2108	.9775	.2156	4.6382		50'
	20'	.2136	.9769	.2186	4.5736		40'
	30'	.2164	.9763	.2217	4.5107		30'
	40'	.2193	.9757	.2247	4.4494		20'
	50'	.2221	.9750	.2278	4.3897		10'
13°	0'	.2250	.9744	.2309	4.3315	77°	0'
	10'	.2278	.9737	.2339	4.2747		50'
	20'	.2306	.9730	.2370	4.2193		40'
	30'	.2334	.9724	.2401	4.1653		30'
	40'	.2363	.9717	.2432	4.1126		20'
	50'	.2391	.9710	.2462	4.0611		10'
14°	0'	.2419	.9703	.2493	4.0108	76°	0'
	10'	.2447	.9696	.2524	3.9617		50'
	20'	.2476	.9689	.2555	3.9136		40'
	30'	.2504	.9681	.2586	3.8667		30'
	40'	.2532	.9674	.2617	3.8208		20'
	50'	.2560	.9667	.2648	3.7760		10'
15°	0'	.2588	.9659	.2679	3.7321	75°	0'
Angle		cos	sin	cot	tan	Angle	

Table V.   Continued.

Angle		sin	cos	tan	cot	Angle	
15°	0'	.2588	.9659	.2679	3.7321	75°	0'
	10'	.2616	.9652	.2711	3.6891		50'
	20'	.2644	.9644	.2742	3.6470		40'
	30'	.2672	.9636	.2773	3.6059		30'
	40'	.2700	.9628	.2805	3.5656		20'
	50'	.2728	.9621	.2836	3.5261		10'
16°	0'	.2756	.9613	.2867	3.4874	74°	0'
	10'	.2784	.9605	.2899	3.4495		50'
	20'	.2812	.9596	.2931	3.4124		40'
	30'	.2840	.9588	.2962	3.3759		30'
	40'	.2868	.9580	.2994	3.3402		20'
	50'	.2896	.9572	.3026	3.3052		10'
17°	0'	.2924	.9563	.3057	3.2709	73°	0'
	10'	.2952	.9555	.3089	3.2371		50'
	20'	.2979	.9546	.3121	3.2041		40'
	30'	.3007	.9537	.3153	3.1716		30'
	40'	.3035	.9528	.3185	3.1397		20'
	50'	.3062	.9520	.3217	3.1084		10'
18°	0'	.3090	.9511	.3249	3.0777	72°	0'
	10'	.3118	.9502	.3281	3.0475		50'
	20'	.3145	.9492	.3314	3.0178		40'
	30'	.3173	.9483	.3346	2.9887		30'
	40'	.3201	.9474	.3378	2.9600		20'
	50'	.3228	.9465	.3411	2.9319		10'
19°	0'	.3256	.9455	.3443	2.9042	71°	0'
	10'	.3283	.9446	.3476	2.8770		50'
	20'	.3311	.9436	.3508	2.8502		40'
	30'	.3338	.9426	.3541	2.8239		30'
	40'	.3365	.9417	.3574	2.7980		20'
	50'	.3393	.9407	.3607	2.7725		10'
20°	0'	.3420	.9397	.3640	2.7475	70°	0'
	10'	.3448	.9387	.3673	2.7228		50'
	20'	.3475	.9377	.3706	2.6985		40'
	30'	.3502	.9367	.3739	2.6746		30'
	40'	.3529	.9356	.3772	2.6511		20'
	50'	.3557	.9346	.3805	2.6279		10'
21°	0'	.3584	.9336	.3839	2.6051	69°	0'
	10'	.3611	.9325	.3872	2.5826		50'
	20'	.3638	.9315	.3906	2.5605		40'
	30'	.3665	.9304	.3939	2.5386		30'
	40'	.3692	.9293	.3973	2.5172		20'
	50'	.3719	.9283	.4006	2.4960		10'
22°	0'	.3746	.9272	.4040	2.4751	68°	0'
	10'	.3773	.9261	.4074	2.4545		50'
	20'	.3800	.9250	.4108	2.4342		40'
	30'	.3827	.9239	.4142	2.4142		30'
Angle		cos	sin	cot	tan	Angle	

Table V. Continued.

Angle		sin	cos	tan	cot	Angle	
	30'	.3827	.9239	.4142	2.4142		30'
	40'	.3854	.9228	.4176	2.3945		20'
	50'	.3881	.9216	.4210	2.3750		10'
23°	0'	.3907	.9205	.4245	2.3559	67°	0'
	10'	.3934	.9194	.4279	2.3369		50'
	20'	.3961	.9182	.4314	2.3183		40'
	30'	.3987	.9171	.4348	2.2998		30'
	40'	.4014	.9159	.4383	2.2817		20'
	50'	.4041	.9147	.4417	2.2637		10'
24°	0'	.4067	.9135	.4452	2.2460	66°	0'
	10'	.4094	.9124	.4487	2.2286		50'
	20'	.4120	.9112	.4522	2.2113		40'
	30'	.4147	.9100	.4557	2.1943		30'
	40'	.4173	.9088	.4592	2.1775		20'
	50'	.4200	.9075	.4628	2.1609		10'
25°	0'	.4226	.9063	.4663	2.1445	65°	0'
	10'	.4253	.9051	.4699	2.1283		50'
	20'	.4279	.9038	.4734	2.1123		40'
	30'	.4305	.9026	.4770	2.0965		30'
	40'	.4331	.9013	.4806	2.0809		20'
	50'	.4358	.9001	.4841	2.0655		10'
26°	0'	.4384	.8988	.4877	2.0503	64°	0'
	10'	.4410	.8975	.4913	2.0353		50'
	20'	.4436	.8962	.4950	2.0204		40'
	30'	.4462	.8949	.4986	2.0057		30'
	40'	.4488	.8936	.5022	1.9912		20'
	50'	.4514	.8923	.5059	1.9768		10'
27°	0'	.4540	.8910	.5095	1.9626	63°	0'
	10'	.4566	.8897	.5132	1.9486		50'
	20'	.4592	.8884	.5169	1.9347		40'
	30'	.4617	.8870	.5206	1.9210		30'
	40'	.4643	.8857	.5243	1.9074		20'
	50'	.4669	.8843	.5280	1.8940		10'
28°	0'	.4695	.8829	.5317	1.8807	62°	0'
	10'	.4720	.8816	.5354	1.8676		50'
	20'	.4746	.8802	.5392	1.8546		40'
	30'	.4772	.8788	.5430	1.8418		30'
	40'	.4797	.8774	.5467	1.8291		20'
	50'	.4823	.8760	.5505	1.8165		10'
29°	0'	.4848	.8746	.5543	1.8040	61°	0'
	10'	.4874	.8732	.5581	1.7917		50'
	20'	.4899	.8718	.5619	1.7796		40'
	30'	.4924	.8704	.5658	1.7675		30'
	40'	.4950	.8689	.5696	1.7556		20'
	50'	.4975	.8675	.5735	1.7437		10'
30°	0'	.5000	.8660	.5774	1.7321	60°	0'
Angle		cos	sin	cot	tan	Angle	

Table V. Continued.

Angle		sin	cos	tan	cot	Angle	
30°	0'	.5000	.8660	.5774	1.7321	60°	0'
	10'	.5025	.8646	.5812	1.7205		50'
	20'	.5050	.8631	.5851	1.7090		40'
	30'	.5075	.8616	.5890	1.6977		30'
	40'	.5100	.8601	.5930	1.6864		20'
	50'	.5125	.8587	.5969	1.6753		10'
31°	0'	.5150	.8572	.6009	1.6643	59°	0'
	10'	.5175	.8557	.6048	1.6534		50'
	20'	.5200	.8542	.6088	1.6426		40'
	30'	.5225	.8526	.6128	1.6319		30'
	40'	.5250	.8511	.6168	1.6212		20'
	50'	.5275	.8496	.6208	1.6107		10'
32°	0'	.5299	.8480	.6249	1.6003	58°	0'
	10'	.5324	.8465	.6289	1.5900		50'
	20'	.5348	.8450	.6330	1.5798		40'
	30'	.5373	.8434	.6371	1.5697		30'
	40'	.5398	.8418	.6412	1.5597		20'
	50'	.5422	.8403	.6453	1.5497		10'
33°	0'	.5446	.8387	.6494	1.5399	57°	0'
	10'	.5471	.8371	.6536	1.5301		50'
	20'	.5495	.8355	.6577	1.5204		40'
	30'	.5519	.8339	.6619	1.5108		30'
	40'	.5544	.8323	.6661	1.5013		20'
	50'	.5568	.8307	.6703	1.4919		10'
34°	0'	.5592	.8290	.6745	1.4826	56°	0'
	10'	.5616	.8274	.6787	1.4733		50'
	20'	.5640	.8258	.6830	1.4641		40'
	30'	.5664	.8241	.6873	1.4550		30'
	40'	.5688	.8225	.6916	1.4460		20'
	50'	.5712	.8208	.6959	1.4370		10'
35°	0'	.5736	.8192	.7002	1.4281	55°	0'
	10'	.5760	.8175	.7046	1.4193		50'
	20'	.5783	.8158	.7089	1.4106		40'
	30'	.5807	.8141	.7133	1.4019		30'
	40'	.5831	.8124	.7177	1.3934		20'
	50'	.5854	.8107	.7221	1.3848		10'
36°	0'	.5878	.8090	.7265	1.3764	54°	0'
	10'	.5901	.8073	.7310	1.3680		50'
	20'	.5925	.8056	.7355	1.3597		40'
	30'	.5948	.8039	.7400	1.3514		30'
	40'	.5972	.8021	.7445	1.3432		20'
	50'	.5995	.8004	.7490	1.3351		10'
37°	0'	.6018	.7986	.7536	1.3270	53°	0'
	10'	.6041	.7969	.7581	1.3190		50'
	20'	.6065	.7951	.7627	1.3111		40'
	30'	.6088	.7934	.7673	1.3032		30'
Angle		cos	sin	cot	tan	Angle	

Table V. Continued.

Angle		sin	cos	tan	cot	Angle	
	30'	.6088	.7934	.7673	1.3032		30'
	40'	.6111	.7916	.7720	1.2954		20'
	50'	.6134	.7898	.7766	1.2876		10'
38°	0'	.6157	.7880	.7813	1.2799	52°	0'
	10'	.6180	.7862	.7860	1.2723		50'
	20'	.6202	.7844	.7907	1.2647		40'
	30'	.6225	.7826	.7954	1.2572		30'
	40'	.6248	.7808	.8002	1.2497		20'
	50'	.6271	.7790	.8050	1.2423		10'
39°	0'	.6293	.7771	.8098	1.2349	51°	0'
	10'	.6316	.7753	.8146	1.2276		50'
	20'	.6338	.7735	.8195	1.2203		40'
	30'	.6361	.7716	.8243	1.2131		30'
	40'	.6383	.7698	.8292	1.2059		20'
	50'	.6406	.7679	.8342	1.1988		10'
40°	0'	.6428	.7660	.8391	1.1918	50°	0'
	10'	.6450	.7642	.8441	1.1847		50'
	20'	.6472	.7623	.8491	1.1778		40'
	30'	.6494	.7604	.8541	1.1708		30'
	40'	.6517	.7585	.8591	1.1640		20'
	50'	.6539	.7566	.8642	1.1571		10'
41°	0'	.6561	.7547	.8693	1.1504	49°	0'
	10'	.6583	.7528	.8744	1.1436		50'
	20'	.6604	.7509	.8796	1.1369		40'
	30'	.6626	.7490	.8847	1.1303		30'
	40'	.6648	.7470	.8899	1.1237		20'
	50'	.6670	.7451	.8952	1.1171		10'
42°	0'	.6691	.7431	.9004	1.1106	48°	0'
	10'	.6713	.7412	.9057	1.1041		50'
	20'	.6734	.7392	.9110	1.0977		40'
	30'	.6756	.7373	.9163	1.0913		30'
	40'	.6777	.7353	.9217	1.0850		20'
	50'	.6799	.7333	.9271	1.0786		10'
43°	0'	.6820	.7314	.9325	1.0724	47°	0'
	10'	.6841	.7294	.9380	1.0661		50'
	20'	.6862	.7274	.9435	1.0599		40'
	30'	.6884	.7254	.9490	1.0538		30'
	40'	.6905	.7234	.9545	1.0477		20'
	50'	.6926	.7214	.9601	1.0416		10'
44°	0'	.6947	.7193	.9657	1.0355	46°	0'
	10'	.6967	.7173	.9713	1.0295		50'
	20'	.6988	.7153	.9770	1.0235		40'
	30'	.7009	.7133	.9827	1.0176		30'
	40'	.7030	.7112	.9884	1.0117		20'
	50'	.7050	.7092	.9942	1.0058		10'
45°	0'	.7071	.7071	1.0000	1.0000	45°	0'
Angle		cos	sin	cot	tan	Angle	

INDEX